Advances in Composite Materials Development

Edited by Dumitra Lucan

Published in London, United Kingdom

IntechOpen

Supporting open minds since 2005

Advances in Composite Materials Development
http://dx.doi.org/10.5772/intechopen.77575
Edited by Dumitra Lucan

Contributors
Elena O. Nasakina, Sevostyanov Michail, Mikhail A. Kaplan, Alexander S. Baikin, Konstantin V. Sergienko, Sergey V. Konushkin, Ilya M. Fedyuk, Alexander V. Leonov, Alexey G. Kolmakov, Dennis Tucker, Vikas Verma, Alexandra Khvan, Rais Ahmad, Ladislav Écsi, Pavel Élesztős, Róbert Jerábek, Roland Jančo, Branislav Hučko

First published in London, United Kingdom, 2019 by IntechOpen
IntechOpen is the global imprint of INTECHOPEN LIMITED, registered in England and Wales, registration number: 11086078, The Shard, 25th floor, 32 London Bridge Street
London, SE19SG - United Kingdom
Printed in Croatia

British Library Cataloguing-in-Publication Data
A catalogue record for this book is available from the British Library

Additional hard and PDF copies can be obtained from orders@intechopen.com

Advances in Composite Materials Development
Edited by Dumitra Lucan
p. cm.
Print ISBN 978-1-78984-129-9
Online ISBN 978-1-78984-130-5
eBook (PDF) ISBN 978-1-78984-624-9

For EU product safety concerns:
IN TECH d.o.o., Prolaz Marije Krucifikse Kozulić 3, 51000 Rijeka, Croatia, info@intechopen.com or visit our website at intechopen.com.

Meet the editor

After graduating as a chemical engineering engineer at the Polytechnic Institute of Bucharest in 1983, Dumitra Lucan joined the Institute for Nuclear Research Pitesti. She obtained her doctor degree from the Polytechnic University of Bucharest in 2003 with a dissertation titled "Contributions on the Study and Mathematical Modelling of the Impurities Concentration and Deposition Processes, by Boiling, into CANDU Steam Generator." Her interests include research and development in the corrosion behavior of the CANDU Steam Generator structural materials depending on material properties, corrosive environment and testing conditions, data processing, and interpretation. She also studied the chemistry of liquid lead and its interaction with structural materials for the Generation IV Lead Fast Reactor.

Contents

Preface

This book explores the research results dedicated to obtaining, characterizing, and mathematically modeling composite materials, especially metal matrix composites (MMCs), with superior properties having a wide range of applications.

The book consists of seven chapters allocated between four sections:

i. Obtaining and characterization

ii. Mechanical and wear behavior

iii. Material's models developing

iv. Applications for nuclear fuel elements

Authors from different countries (India, Russia, Slovakia, and the United States) have contributed chapters to this book.

Chapter 1 focuses on the investigation of biocompatible composites obtained using magnetron sputtering for the production of minimally invasive medical implantation devices (stents). Nano- and microdimensional surface layers of Ta, Ti, Ag, and Cu on flat and wire NiTi, Cu, Ti, and SiO_2 substrates are created. Phase composition, surface morphology, and layer-by-layer composition are investigated on an X-ray diffractometer, scanning electron microscope (SEM), and Auger spectrometer. It is shown that the thickness and structure of surface layers are affected by sputtering distance, time, power, and bias voltage at the substrate. The presence of the transition layer that contains both substrate and target elements and provides high adhesion of the surface layer to the substrate is demonstrated. The material is tested for corrosion resistance under static conditions by dipping into solutions with various acidities (from pH 1.68 to pH 9.18) for two years, static mechanical properties and biocompatibility in vitro and in vivo. A slight corrosive dissolution is observed only in a medium at a pH of 1.56. Dissolution in the other media is absent. An increase in strength and plasticity in comparison with substrate is attained depending on the nature of the sputtered substance and substrate. The toxicity of samples is not revealed.

Chapter 2 contains experimental results on the synthesis and characterization of a polyaniline zinc oxide nanocomposite (PAZO) by different analytical techniques such as FT-IR, XTD, TGA-DTG, SEM, and TEM. Nanocomposite material is further explored for the removal of Cr(VI) from aqueous solution, and the effect of various adsorption parameters such as agitation time, solution pH, adsorbent dose, initial metal ion concentration, and temperature is observed and optimized by preliminary experiments. The adsorption of metal ions is highly pH dependent and the maximum removal efficiencies and adsorption capacities of the selected metal ions are obtained at pH 2. The experimental data are tested using Langmuir, Freundlich, D-R, and Temkin models and the data are best followed by the Langmuir model. The maximum monolayer adsorption capacity is 120.92 mg g^{-1} at 30°C, 134.22

mg g^{-1} at 40°C, and 139.47 mg g^{-1} at 50°C. All kinetic parameters suggest that the adsorption of metal ions by PAZO follows second-order kinetics and chemisorption is the rate-determining step. The positive values of $\Delta H°$ and negative value of $\Delta G°$ indicate that the adsorption process is endothermic and spontaneous in nature.

Chapter 3 highlights the fact that Al and its composites are the best-suited materials and have better properties than unreinforced materials. Beneficial properties with reduced prices have enlarged their applications. The chapter consists of a literature survey on Al-based MMCs and especially their physical, mechanical, and wear characteristics. These unique properties (high hardness, high strength, high stiffness, high wear, and abrasion and corrosion resistance) make it possible for Al MMC to be used successfully in defense, aerospace, automotive, aviation, and thermal management areas in engine pistons, cylinder barrels, connecting rods, and elements of vehicle braking systems. Physical and mechanical properties and wear behavior of Al-based metal matrix composites are two subsections of the chapter. One of the examples shown in the chapter consists of the Ni contribution to the wear of Al MMCs. Maximum wear resistance is obtained with the addition of 3 wt% Ni to pure Al under both loads and against both counterfaces. The wear resistance of the alloys increases with increasing Ni content up to 3 wt% Ni and tends to decrease to less than 3 wt% Ni. The wear rate of the Al–xNi alloys increases with increasing applied normal load. Many such examples are presented in this chapter. The work is valuable because it presents many ways of improving the properties of Al MMCs.

Chapter 4 presents an alternative framework for developing objective and thermodynamically consistent hypoelastic-plastic and hyperelastic-plastic-based material models using the first nonlinear continuum theory of finite deformations of elastoplastic media. The theoretical study contains the following steps: a short overview of the nonlinear continuum mechanical theory for finite deformations of elastoplastic media; modeling of plastic flow in the material; constitutive equations of the material; reference definition of the yield surface; calculation of the plastic multiplier; and the ratio of ductile and total damage increment. Furthermore, the chapter presents numerical and experimental results. The related material models are demonstrated in numerical experiments. The most important implication of the presented theory is that the analysis results of the related models are no longer affected by the description and particularities of the mathematical formulation. Nonlinear continuum theory is also briefly presented, while thermodynamic consistency of the formulation is discussed in detail. Another important implication of the theory is that the dissipated plastic power density of the model can be directly related to the dissipated plastic power density of the specimen coming from the uniaxial tensile stress of the modeled material. Multiscale analyses and multiobjective optimization of thermomechanical properties of composite materials, particularly in a nonlinear regime, still need relevant and computationally efficient models. The models presented in this chapter could be useful in the abovementioned field.

Chapter 5 is focused on specific research with the goal of producing CERMET fuels for nuclear thermal propulsion (NTP) as an alternative to chemical propulsion. Initially, a brief history of NTP by NASA, including fuel element work, followed by more recent research on various fuel systems under consideration is presented. At present, there are a number of fuels under consideration for NTP. These include graphite composites, tricarbides (U-Zr-Nb)C, and CERMETS (MUO_2 and W/UO_2, Mo/UO_2, W/UN, and Mo/UN). For W/UO_2, the loss of uranium from the UO_2 particles and subsequent migration into the tungsten matrix can be understood in terms of the generation of oxygen vacancies during sintering in a vacuum environment.

It is found that various oxides such as ThO_2, Ce_2O_3, and Y_2O_3 when added to the CERMET powder reduce fuel loss. It is also found that the oxide additives do not increase the solubility of uranium in UO_2, but stabilize UO_2 against oxygen loss. Two mechanisms are proposed to explain this stabilization: (1) oxide additives lower the partial molar free energy of oxygen in the UO_2 without the possibility of forming free uranium upon cooling and (2) with the addition of metal oxide, uranium is transformed to a hexavalent state, which does not reduce to uranium metal. In conclusion, the most likely candidate to stabilize UO_2 during sintering and thermal cycling in hydrogen will be the addition of a rare earth oxide.

The book is intended for practical engineers, researchers, students, and others dealing with the reviewed problems. We hope that the book will be beneficial to all readers and initiate further inquiries and developments in the field of advanced materials with aspirations for a better future.

On behalf of all the authors, I want to express our gratitude to publishing author service manager Mr. Luka Cvjetkovic for his patience and understanding.

Dr. Dumitra Lucan
Technologies for Nuclear Energy State Owned Company,
Institute for Nuclear Research, Nuclear Materials and Corrosion Department,
Pitesti, Romania

Technical Sciences Academy of Romania ASTR,
Bucharest, Romania

Section 1

Obtaining and Characterization

Chapter 1

Using of Magnetron Sputtering for Biocompatible Composites Creating

Elena O. Nasakina, Mikhail A. Sevostyanov, Alexander S. Baikin, Sergey V. Konushkin, Konstantin V. Sergienko, Mikhail A. Kaplan, Ilya M. Fedyuk, Alexander V. Leonov and Alexey G. Kolmakov

Abstract

Biocompatible composites obtained using the magnetron sputtering for the production of minimally invasive implantation medical devices (stents) were investigated. Nano- and microdimensional surface layers of Ta, Ti, Ag, and Cu on flat and wire NiTi, Cu, Ti, and SiO_2 substrates were created. The phase composition, surface morphology, and the layer-by-layer composition were investigated on an X-ray diffractometer, SEM, and Auger spectrometer. It was shown that the thickness and the structure of surface layers were affected by the sputtering distance, time, power, and the bias voltage at the substrate. The presence of the transition layer that contains both substrate and target elements and provides high adhesion of the surface layer to the substrate has been demonstrated. The material was tested for corrosion resistance under static conditions by dipping into solutions with various acidities (pH from 1.68 to 9.18) for 2 years, static mechanical properties, and biocompatibility in vitro and in vivo. A slight corrosive dissolution was observed only in a medium with a pH of 1.56. Dissolution in the other media is absent. An increase in strength and plasticity in comparison with substrate was attained depending on the nature of the sputtered substance and substrate. Toxicity of samples has not been revealed.

Keywords: surface layer, magnetron sputtering, biocompatibility, corrosion resistance, shape memory effect and pseudoelasticity

1. Introduction

In the modern world, an efficient manner of operational characteristic increase and of classical material shortcoming elimination is a formation on their basis of composite materials [1–6]. Development of the layered composite materials allowing to effectively combine desirable operational characteristics of the modified surface layers and the main material (substrate) at the present time is perspective in many areas of human activity: in optics (conducting, antireflecting, filtering, reflecting, and absorbing media), electronics (conductors, semiconductors, dielectrics), machine engineering, building and household (tribological, durable, wear-resistant, functional, protective, resistant to action of aggressive environment, decorative, and other coatings for structural and utility materials), medicine (biocompatible materials), etc. [1–17].

An effective and fairly common method for such surfaces formation is physical vapor deposition, including variations of magnetron sputtering, since at relatively small expenditure of time and resources, it allows to efficiently receive qualitative thin films of a diverse nature on substrate of virtually any nature and geometry and to control properties of the created materials [7–17]. The deposition method developed at the Baikov Institute of Metallurgy and Materials Science allows one to produce [coating—transition layer—substrate] nanocompositions from thermodynamically nonmiscible elements with good adhesion and resistance to external thermal and mechanical influences [18] and at the same time to avoid substrate overheating with bombarding electrons by keeping them near the sputtering target, which is of high importance for substrate materials with low melting points or phase structure that is sensitive to temperature changes [7].

Among other things, this technology can be successfully used in the formation of medical composite materials that need to have complex properties, combining only the required characteristics of classical materials—for example, for the production of noninvasive stent implants designed to restore the patency of hollow parts of the cardiovascular, excretory, digestive, and respiratory systems [19, 20]. At the same time, the parameters of the resulting composites depend on a number of process conditions that can be conveniently varied over a wide range.

In addition, the initial choice of the components of the future composite is important. Materials with the shape memory effect are the best candidates for creation of medical implants plastically deformable in the cooled condition to extremely compact type promoting easier and less traumatic delivery to the necessary site of an organism without serious surgical intervention. Then, they independently take the functional form in set operational conditions without additional effect [20–23]. The most known medical material from this class is nitinol (NiTi) endowed with mechanical characteristics similar to behavior of living tissues that helps it to adapt to physiological loadings providing necessary service conditions [20–23]. But in addition to positive mechanical characteristics, this alloy possesses also a number of shortcomings: difficulty of processing in case of product production, the high content of a toxic element, disputable level of biocompatibility, and corrosion resistance. Research toward its improvement is actively conducted [23–25]. Thus, nitinol can be taken as the basis of the composite (substrate) and a biologically inert barrier at the contact between the metallic parts of the implant and active biological body fluids is needed. A new surface should be represented by a material with high corrosion resistance and biocompatibility. For example, tantalum and titanium is interesting due to high corrosion resistance in aggressive media, radiopacity, conductivity, wear resistance, nontoxicity, etc. [6]. Silver also exhibiting antitumor and antibacterial action is one of the suitable materials [26].

The purpose of this work was to investigate capabilities and regularities of production of layered biomedical composite materials based on an NiTi shape memory alloy with a surface layer from highly corrosion-resistant and biocompatible tantalum, titanium, or silver with strong adhesion between the components formed by magnetron sputtering and to study its operational properties.

2. Obtaining and investigating of biocompatible composites of medical appointment

In this work, creation of layered composites was carried out by formation of tantalum, titanium, silver, etc., surface layers on flat and wire nitinol, titanium, glass/SiO_2, etc., substrates (basis) by magnetron sputtering in an argon atmosphere by using a Torr International facility (United States). Working and residual pressure in the vacuum chamber was 0.4 and 4×10^{-4} Pa, respectively.

The surface layers were produced under the following conditions of the process: (1) direct current magnetron, in the case of tantalum and titanium layers I ~ 400–1100 mA, U ~ 360–700 V; in the case of silver layers I ≈ 865 mA, U ~ 830 V; (2) with substrate rotation (rate was 9 rpm) and without it; (3) sputtering time t = 5 to 120 min; (4) bias voltage Ub ≈ 0–1000 V; and (5) sputtering distance (the distance from the target to the substrate) of 40–200 mm.

To avoid overheating, the substrate is critically important for substrate materials with phase structures that are sensitive to temperature changes as nitinol, for example—the thermal treatment allows one to vary static properties and cyclic loadings in operating conditions with a wide range of deformations and is extremely important for stabilization of the properties, constraining (shaping) the samples, and successful application of the product.

To determine the substrate surface temperature, we used special control samples from materials with various melting points: In (tm = 156.4°C), Sn (tm = 231.9°C), Pb (tm = 327.4°C, and Zn (tm = 419.5°C). Since none of the metals showed surface melting, we concluded that, under any conditions, the substrate surface temperature did not reach 150°C in any of the regimes used.

Disks made from chemically pure tantalum, titanium, silver, copper, etc., were used as the sputtered target. Plates made from nitinol, titanium, copper, steel, glass/SiO_2, etc., with a size of 10 × 10 × 0.5 mm and 280 μm diameter wires from nanostructured nitinol (with the composition 55.91 wt% Ni + 44.03 wt% Ti, having grains in the form of 30 to 70 nm diameter nanowires and a cubic crystal lattice (B2 phase)) were used as the basis for composites. Plates were treated with abrasive sandpaper (from 400 to 800 grit) and polished (until their surface became mirror-like) with the addition of diamond suspensions with a particle size of 3, 1, and 0.05 μm for the removal of flat indentations and defects. Nitinol wires were also polished consecutively with sandpaper from 180 to 1000 grit and subjected to finishing polishing with GOI (State Optical Institute) paste to a mirror surface. The decrease in the diameter was to 10 μm in comparison with the original. The depth of surface defects after the processing was less than 1 μm. Different substrates were selected to perfect the production of layered composites. Silicon, copper, and steel substrates are also of interest as a basis for production of functional materials for a wide range of applications (optics, electronics, construction materials, etc.). To clean, activate, and polish the substrate surface, before sputtering, it was bombarded with argon ions at Ue = 900 V and Ie = 80 mA; i.e., preliminary ionic etching (PIE) was performed.

Phase structure of the deposited films was characterized by the Ultima IV X-ray diffractometer (Rigaku Co., Japan) in Cu Kα—radiation on the base of Bragg—Brentano method. Phase analysis was prepared in the PDXL program complex using the ICDD database. The surface morphology and the layer-by-layer composition were investigated on a scanning election microscope (SEM) VEGA II SBU with the module INCA Energy for energy-dispersive analysis (TESCAN, Czech Republic), on a GDS 850A atomic emission spectrometer (JEOL Co., Japan), and on a JAMP-9500F Auger spectrometer (JEOL Co., Japan) in combination with ion etching at argon bombardment under an angle of 30°. In Auger electron spectroscopy, the surface layer thickness was taken to be equal to the depth at which the atomic percentages of the constituent elements plateaued. Fracture surfaces were also examined on the TESCAN VEGA II SBU.

On the whole, similar results were obtained by examining the surface layer composition (**Figure 1**): the top surface layer was oxygen-enriched to a depth of 20 nm owing to active surface adsorption, a deeper layer consisted of only submitted element, and the transition layer was also oxygen enriched and resided on the substrate. The formation of the transition layer is connected with the fact that magnetron sputtering results in atoms and ions of the sputtered substance not only condensing on the substrate surface but also approaching it with some excess energy with their contact leading to a number of interparticle interactions: "knocking in" of sputtered atoms and ions, "knocking out" by them (upon elastic or inelastic interaction with or without transfer of their additional energy), and redeposition or, on the contrary, penetration of the surface particles (both of the substrate and earlier deposited elements) into the substrate structure, formation of radiation defects that stimulate mutual diffusion of elements of the deposited layer and substrate at their interface, etc. Thus, the mobilized particles, sputtered substance, and surface region of the substrate that are subjected to multiple collisions and set in chaotic motion at or near the substrate surface are constantly mixed. In the end, the surface region becomes so saturated by the sputtered substance that its interaction with new flows of atoms and ions leads to the formation of the surface layer of the composite.

Figure 2a shows dependence of tantalum surface layer thickness from time of magnetron sputtering on a nitinol substrate. Increasing the sputtering time to 20 min (at direct current of 865 mA, voltage of 700 V and distance 200 mm) increased the thickness of both the surface layer (consisting only of the deposited substance) and the transition layer (containing elements of both the substrate and deposited substance). Further increasing the sputtering time caused an increase only in the thickness of the surface layer, which varied nonlinearly, following a descending law, and up to 30 min more intensively, than at bigger time. This situation remains at all used materials and conditions. It occurs, most likely, because at the beginning of surface layer formation, atoms and ions of deposited substance, overcoming a sputtering distance, collide with particles of working gas, with each other and with new substrate surface and do not appear at each site of its surface in equal volume and at the beginning interact with it chaotically and unevenly. And further (at increase of sputtering time, and so of the time of influence on a surface), particles continue to collide, mixing up, try to reach thermodynamically more advantageous position and state, and a more uniform distribution of the deposited substance at the surfaces takes place. By consideration of cross section of samples, it

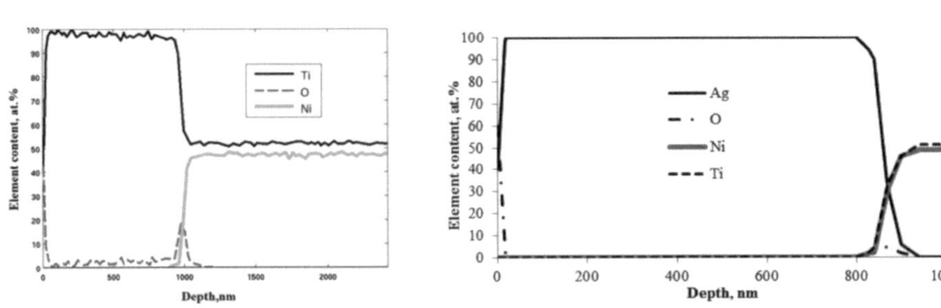

Figure 1.
Composition depth profiles for a Ti-nitinol sample obtained by sputtering for 30 min at direct current of 865 mA, voltage of 700 V, and distance of 200 mm (a) and for Ag-nitinol obtained at a distance of 150 mm (b).

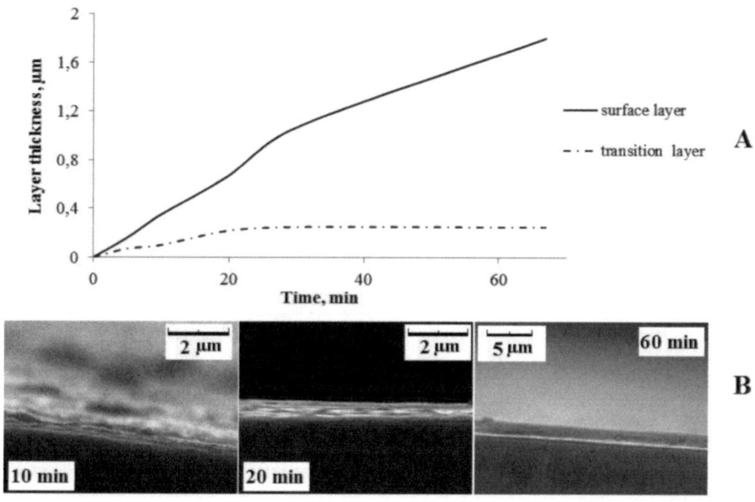

Figure 2.
Dependence of the surface layer structure on the time of magnetron sputtering of tantalum on a nitinol substrate at direct current of 865 mA, voltage of 700 V, and distance 200 mm (a) and microstructure changes (b).

is visible that at smaller time of a sputtering, the surface shows a big heterogeneity (**Figure 2b**). First, the layer had the form of isolated islands. Subsequently, a more uniform Ta distribution over the substrate surface was obtained.

In case of thin films of tantalum, according to literary data, formation of both alpha and beta phases, which differ in properties, is possible [10–17, 27–29]. The X-ray diffraction patterns of our samples with nitinol basis (**Figure 3**) demonstrate that depending on sputtering, time tantalum is formed in two various crystal states—an alpha (a cubic crystal lattice) and a beta (a tetragonal lattice with the small content of oxygen).

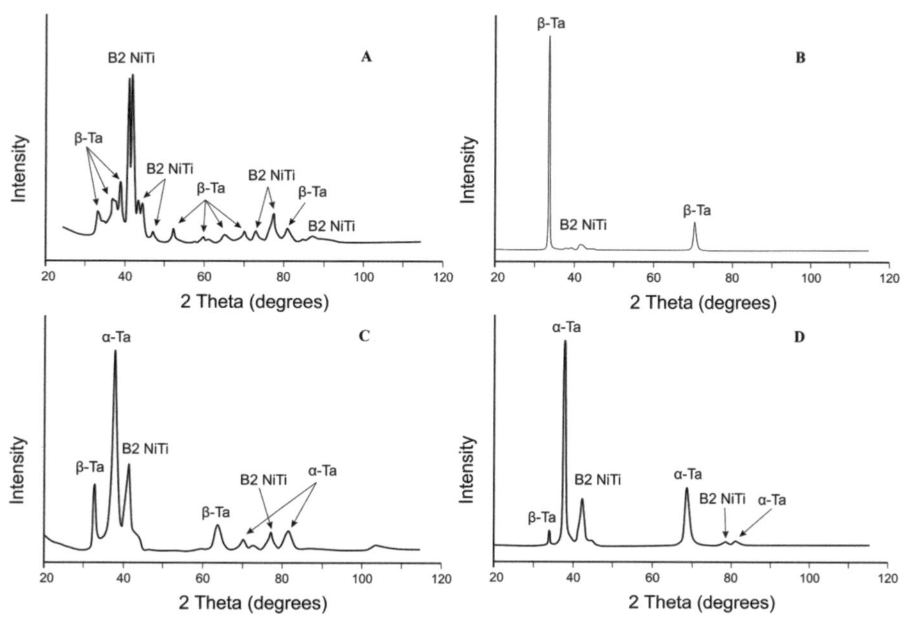

Figure 3.
X-ray diffraction patterns of a composite on the basis of a nitinol received at magnetron sputtering time: (a) 5 min, (b) 10 and 20 min, (c) 29 min, (d) 30 min at direct current of 865 mA, voltage of 700 V and distance of 200 mm.

In the case of a sample with a Ta surface layer obtained by sputtering for 5 min, the major phase was nitinol, but many peaks β-Ta in the 2-theta range from 33 to 81° were observed, which also corresponds to various crystal orientation. In the composites formed after sputtering for 10 and 20 min, β-Ta with O were a major phase, and only two main peaks were observed, but nitinol was also present. After 29 min, the strongest peak was that from α-Ta, and there were β-Ta with O and nitinol, due to the averaging of results over the entire probing depth; at further increase in time, α-Ta dominated and very weak peaks of β-Ta and nitinol were observed. Thus, it turns out that irrespective of summary sputtering time, the beta phase is formed in the beginning and at sputtering time, more than 20 min on it, alpha tantalum is deposited. The same regularities are observed in case of other substrates, which are united by availability of oxygen in a surface. In contrast to data available in the literature, the formation of α-Ta in this study cannot be due to an increase in temperature [10, 14, 27, 28].

Several theories of tantalum formation in α or β phase is developed, which are generally connected with working temperature and pressure (defining mobility and energy of atoms) and the substrate nature. However, different authors achieve often contradictory results.

It is noted that the alpha phase is formed at temperatures more than 400°C, promoting mobility increase in deposited atoms : initially at heating of a substrate or as a result of the annealing following sedimentation (then deposited β-Ta transforms in α-Ta) [10, 14, 27, 28]. However at a temperature about 400–500°C, β phase is also received (for example, in the form of the particles distributed in α) [10, 28], and α is also formed without heating [12, 15]. It is specified that with the growth of temperature, the size of grains, impurity amount in a surface layer (for example, the dissociation of oxides enhanced, i.e., the O contents lowered), and its amorphousness decreases.

Presence at the working atmosphere of the high oxygen content according to [16] leads to fast formation of oxides and, therefore, promotes formation of a tantalic layer in a beta state, whereas in [15], oxygen environment did not prevent the formation of alpha tantalum. At a deposition on silicon and glass substrates in [13, 16, 17], 0.5–0.7 Pa sputtering pressure led to α-Ta formation and smaller or bigger pressure—β-Ta, but in [15] already at 0.28 Pa, α phase was formed.

In [17], alpha tantalum was also formed at 0.3 and 1.4 Pa pressure, but at sputtering on earlier deposited α-Ta (110) layer. Also, it was specified that (110) is the most low-energy lattice for body-centered cubic (BCC) materials and provokes formation on itself of the same structure. Being a zone of a new surface nucleation, the substrate surface specifies the character of its structure formation. It was shown that on amorphous carboniferous or oxidic surfaces, the beta tantalum is formed, for example, on titanium without natural oxide or TaN substrates, the α-Ta is formed [11, 12, 14, 17].

And though availability of oxygen on a substrate surface not always prevented the formation of α-Ta, nevertheless, it is considered that it promotes formation of β-Ta. Therefore, its creation in this work in an initial time period on all substrates is quite expected, despite ionic etching.

In [14], as well as in this research, it was shown that longer time promotes layering of α-Ta on earlier formed β-Ta, but authors connected it with a considerable warming up of a surface (more than 350°C), whereas in this work, temperature of a substrate did not rise higher than 150°C and so could not influence formation of alpha phase. It is worth noting that α-Ta is a more thermodynamically stable phase. It is, therefore, reasonable to assume that, in this study, it results from a more uniform surface coverage with increasing sputtering time (because increasing the sputtering time increases the probability that a particle will find a more appropriate state and position), possible local surface heating (within several atomic layers, which cannot be detected visually), and the absence of oxygen (incorporated into the β-Ta sublayer) [11, 12, 14, 17].

In combination with the above results of layer composition, this X-ray diffraction data lead us to assume that the surfaces of both the substrate and surface layer actively adsorbed oxygen and that, in the initial stage of the process, the sputtering time was too short for a purely metallic tantalum layer to form.

Purely silver or titanium layers in a single phase are formed on nitinol at all conditions, which is reflected in an X-ray pattern by characteristic peaks (**Figure 4**) [26].

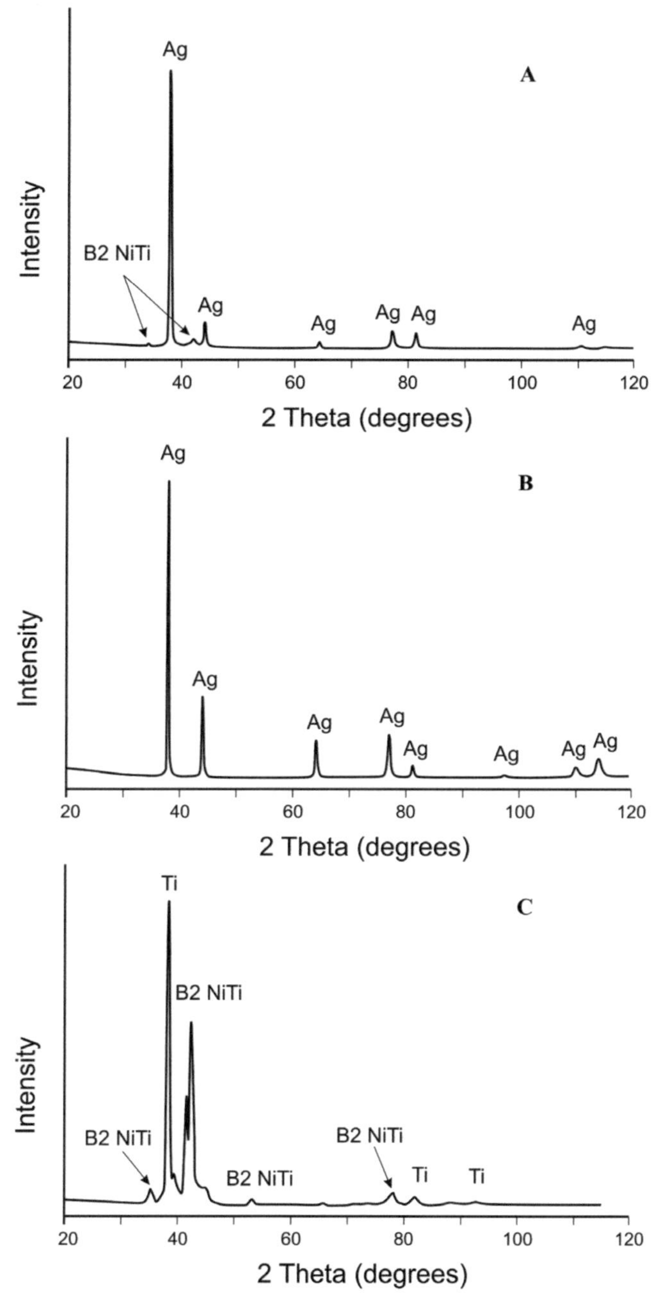

Figure 4.
X-ray patterns of Ag-NiTi composites obtained by sputtering: (a) for 20 min, (b) for 30 min, and (c) Ti-NiTi for 30 min at a distance of 150 mm, 865 mA, and 400 V.

A composite with silver that is produced for 30 min has only Ag peaks observed because of natural growth of the thickness of the surface layer. At a sputtering time of 20 min or in Ti-NiTi composite produced for 20 and 30 min, the main phase is a sputtered metal, but also traces of the nitinol substrate are observed owing to data averaging over the depth. These results are repeated with all the used substrates.

Figure 5 shows the dependence of the thickness of the surface tantalum layer (produced for 30 min at ~865 mA, ~700 V, a sputtering distance of 200 mm and with PIE on a flat nitinol substrate) on the applied negative bias voltage that was instrumental in the process of ion-atomic deposition [7, 18]. The bias voltage affected both the thickness and the structure of layers: a voltage of 100 V reduced (relative to zero voltage bias) the thickness of the surface and the complete layers supposedly because of the structure densification by the additional ionic bombardment. At higher voltages, the surface layer became thicker owing to an increase in the rate of deposition of the sputtered material, while the thickness of the transition layer was reduced somewhat (apparently because of further densification of its structure). The optimum conditions were attained at 500 V, and the further increase in the bias voltage probably led to certain sputtering of the surface: the thickness of the surface layer was reduced again, while the thickness of the transition layer remained the same.

Large corrugations with a length on the order of 20–10 μm and a width of 3–5 μm were observed at the surface of all samples (**Figure 6**). Their characteristic appearance did not change with applied bias voltage, and they did not disappear after a layer with a thickness of about 10 nm was etched out. Therefore, they may be interpreted as the initial microrelief of the sample surface. When the bias voltage was applied, smaller wavelike corrugations with a length of 6–10 μm and a width of 0.5–1 μm emerged at the surfaces of samples. An increase in the bias voltage from 100 to 1000 V led to the gradual smoothing of these corrugations, which were easily observable at first, but became virtually imperceptible at 1000 V. The surface after 1000 V did not differ from the one obtained under zero voltage bias (**Figure 6**). The emergence of smaller scale corrugations may probably be attributed to the presence of residual compressive stresses at the composite surface. A bias voltage of 500 V triggered a uniform distribution of point dimples that likely represented the traces of ion bombardment [9], but these pits and all other inhomogeneities of a similar scale disappeared after a layer with a thickness of about 10 nm was etched out.

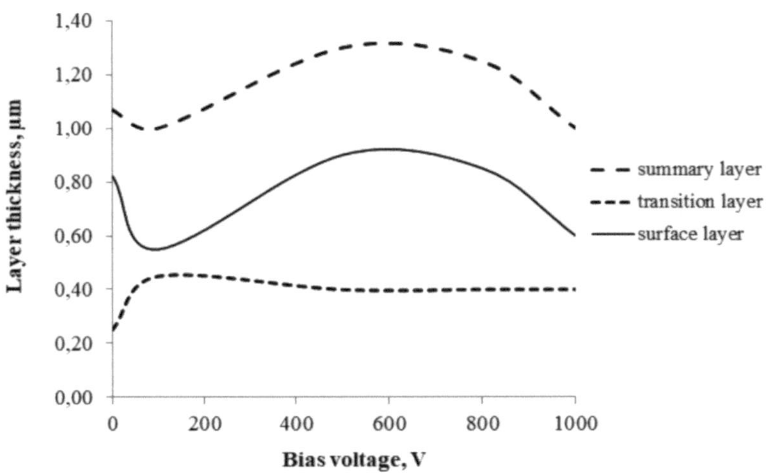

Figure 5.
Dependence of the thickness of the surface tantalum layer produced on a nitinol substrate at ~865 mA, ~700 V, a sputtering distance of 20 cm, and with PIE for 30 min on the bias voltage.

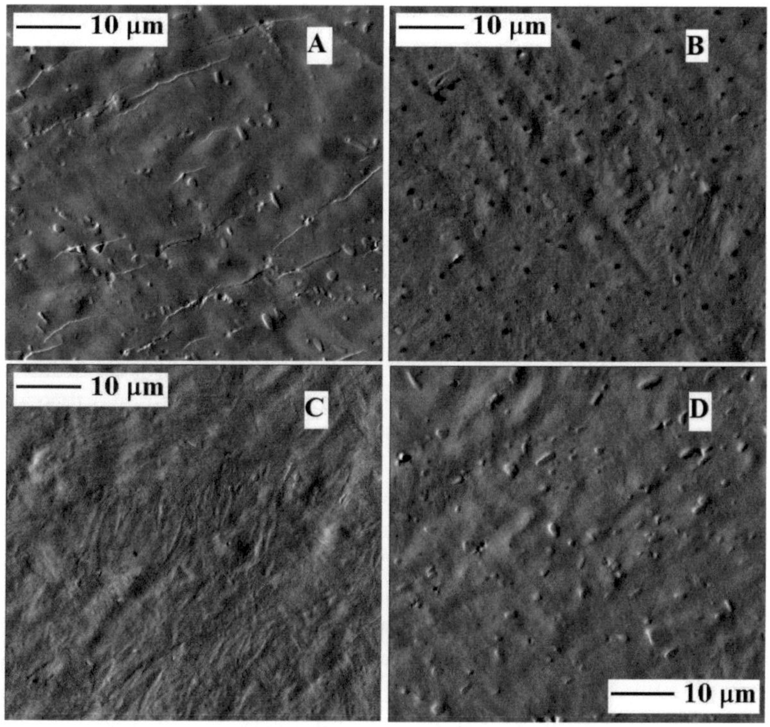

Figure 6.
Appearance of the surface tantalum layer produced on a titanium nickelide substrate at ~865 mA, ~700 V, and a sputtering distance of 15 cm (the sputtering process took 30 min) under Ub = (a) 100, (b) 500, (c) 800, and (d) 1000 V.

The overall thickness of surface layers was increased almost linearly with sputtering power. The thicknesses of both the surface and the transition layers were raised to 30% of the maximum value (**Figure 7**). The thickness of the transition layer was reduced in the 30–50% power interval, while both the surface and the transition layers became thicker again at higher powers. The increase in their thicknesses may be attributed to the raised target sputtering rate [7], and the temporary reduction in the transition layers thickness may be associated with the structure

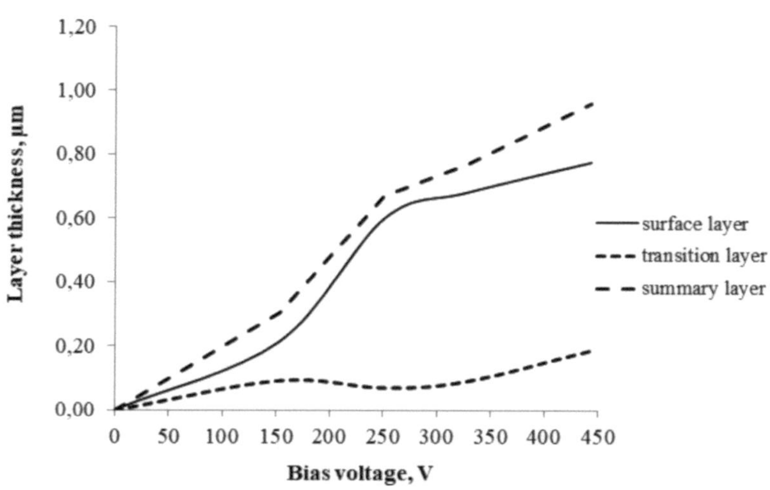

Figure 7.
Dependence of the thickness of the surface titanium layer produced on a glass substrate at ~865 mA, ~400 V, a sputtering distance of 150 mm, Ub ~ 0 V, for 30 min on the sputtering power.

densification and a reduction in the available formation time with increasing energy and density of the flux of sputtered material. When the tantalum sputtering power was raised above 70%, the thickness of the surface layer increased only slightly, and the thickness of the transition layer remained the same (on a metal substrate) or was increased (on a glass substrate; see **Figure 7**) presumably owing to the influence of pores in the material. This also raised the target consumption rate and the potential to contaminate the surface of the composite with, among other things, elements of the walls of the working chamber that are knocked out by high energy particles.

On the one hand, with distance increasing at other equal conditions, the thickness of the tantalum surface layer naturally decreases (**Figure 8**) because larger volume of the sputtered substance is scattered away from the substrate; on the other hand, the thickness of the transition layer increases, which can be explained by a more intense flow of the sputtered substance at a shorter distance, uniformly but faster filling the surface and less diffusing into the substrate; and the total thickness of the layers eventually reaches a certain plateau, practically unchanged when the distance is more than 80–90 mm. Since the presence of a substantial transition layer is a presumable reason for the good adhesion of a new surface to the substrate [6], the surface layer must be adjusted to the mechanical properties of the substrate, and also considering the microdefects of the surface at small distances, the distances from the target to the substrate in the range of 100–150 mm are more optimum.

In contrast to tantalum, where the appearance of the surface layer thickness curve completely corresponds to the calculated models [7], in the case of a silver layer, two plateaus are observed (**Figure 9**): almost constant thicknesses at small sputtering distances can be attributed to saturation of the surface at high intensities of flux falling on the substrate. Unlike the previous case, both the surface layer and the transition layer are thinned with increasing distance, because a smaller volume of substance reached its goal, but the intensity of the flow did not affect the formation of the transition layer.

Visually, the layer thickness also was reduced (**Figure 10**) as the sputtering distance was increased under otherwise equal conditions. At smaller distances from the target to the substrate, no appreciable transition layer was observed in SEM images.

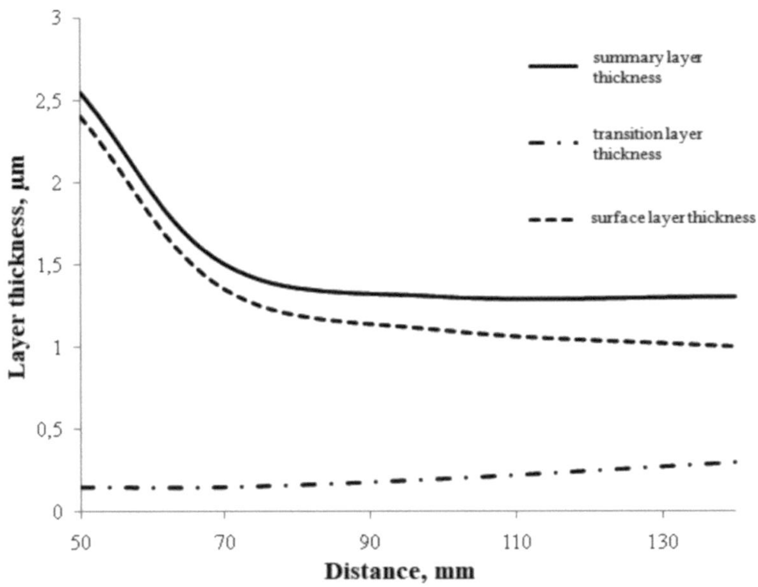

Figure 8.
Change in the tantalum surface layers thickness as a function of the distance between the target and the substrate for composite obtained for 30 min at 400 V, 865 mA.

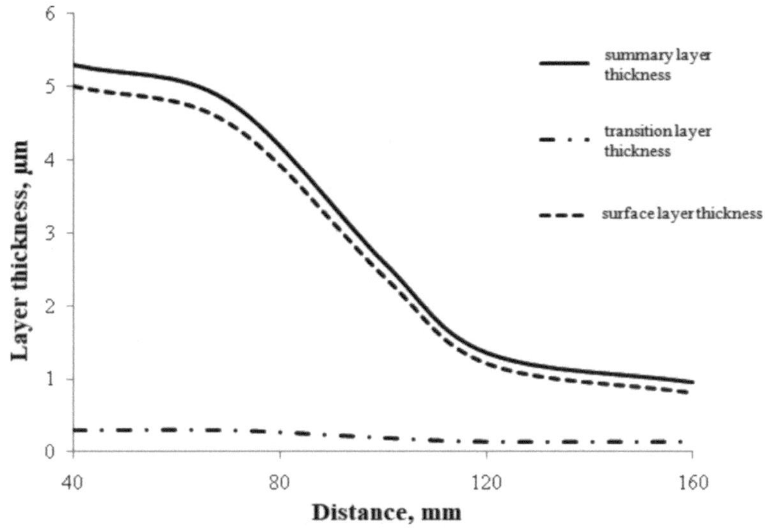

Figure 9.
Change in the silver surface layers thickness as a function of the distance between the target and the substrate for composite obtained for 30 min.

The morphology of a new surface repeats the substrate state regardless of the sputtering conditions (**Figure 11**). However, at small distances, surface microdefects appear as point depressions (**Figure 12**), recalling effect of high bias voltage and ion implantation [7, 9], which is correlated with a more intense flow of the sprayed material reaching the surface of the substrate, in comparison with larger distances. Formation of layers on the side that is opposite to the sputtered flow was noted. In this case, the structure and patterns of these layers changes are analogous to the straight side, but every 10–15 times thinner (**Figure 13**). This also could be accounted for by a large sputtering distance: when particles that are sputtered travel over large distances, they completely lose additional energy and directed movement, slow down to thermal velocities corresponding to the gas temperature, start to move like any atoms in a gaseous state, and can condense at the opposite side of the substrate upon collision with it [7]; there is also the possibility that sputtered atoms

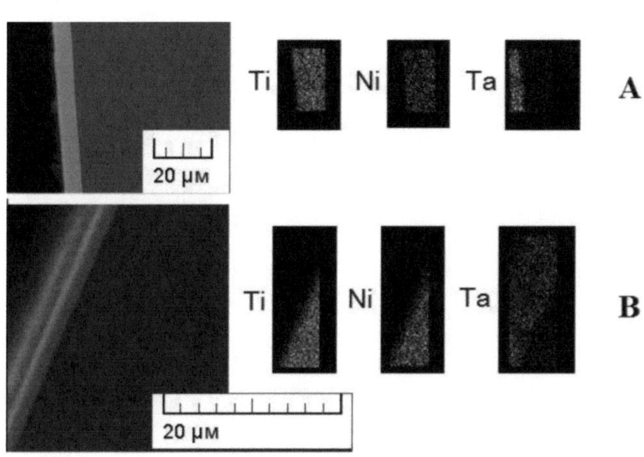

Figure 10.
Dependence of the thickness of the surface tantalum layer distribution of elements in the structure of composites produced on a nitinol substrate for 120 min at ~865 mA, ~700 V, Ub ~ 1000 V, and with PIE on the sputtering distance. The sputtering distance was (a) 100 and (b) 200 mm, while the layer thickness was (a) 6.3 and (b) 3.6 µm.

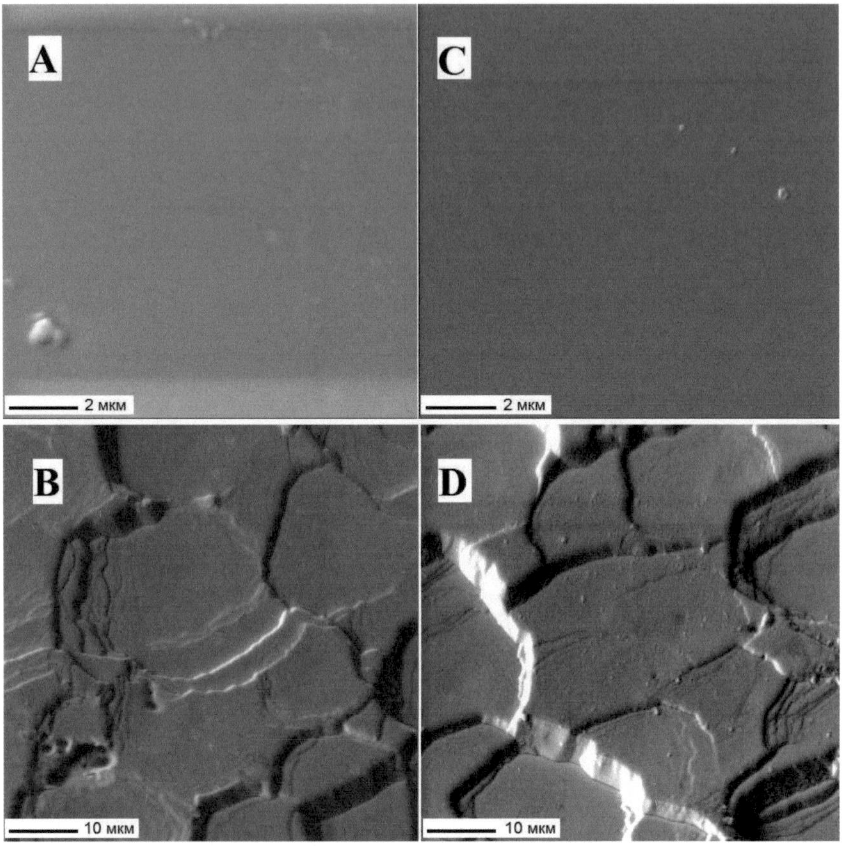

Figure 11.
Morphology of the surface of a glass substrate, straight side (a) and opposite side (b), and a silver layer that is formed on it, straight side (c) and opposite side (d), by sputtering for 20 min at a distance of 70 mm.

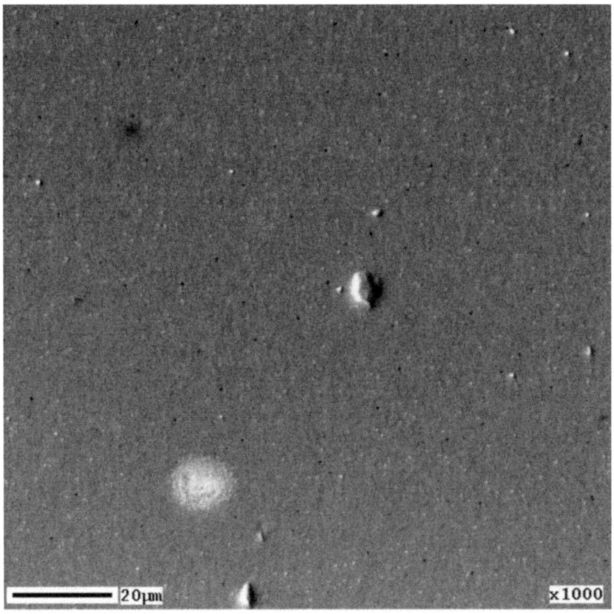

Figure 12.
Morphology of the silver surface layer sputtered for 20 min at a distance of 40 mm.

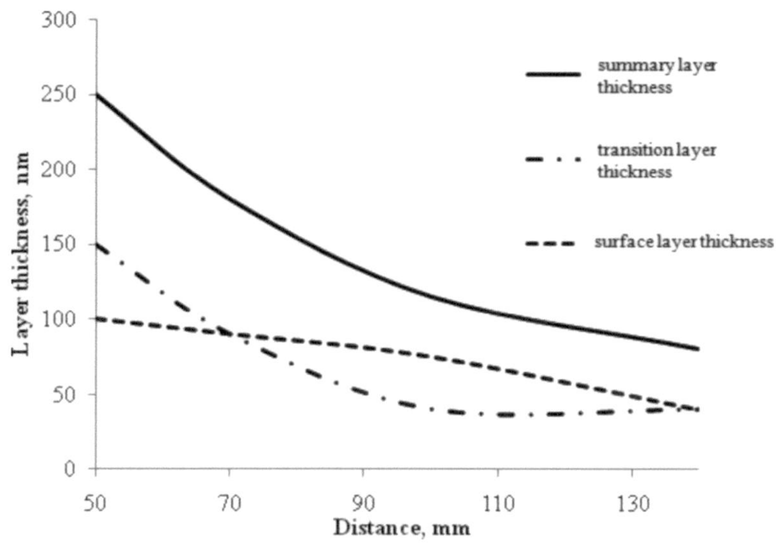

Figure 13.
Change in the tantalum surface layers' thickness as a function of the distance between the target and the substrate on the opposite side of the substrate for composite obtained for 30 min at 400 V, 865 mA.

without complete loss of their kinetic energy fall to the opposite side of the substrate as a result of multiple collisions and reflections on atoms of the working gas.

The study of the composites after static breakdown (metal wire) or brittle fracture (glass) showed that their components (the surface layer and the base) were not separated from each other even in the area of failure (**Figure 14**). It was assumed that the presence of the transition layer was the reason for good adhesion between the surface layer and the base. Preliminary ion etching improved adhesion.

The mechanical properties of samples with a working part length of 45 mm were determined under the conditions of static stretching on an Instron 3382 (Instron, USA) universal testing machine with a loading speed of 2 mm/min. The base diameter was used in the calculation of strength properties. Three to five samples were tested per one experimental point. Micro-Vickers hardness measurements determined at loading 1–2 N by the WOLPERT GROUP 401/402 device—MVD (WILSON Instruments, USA) equipped with a light microscope. The conventional yield strength $\sigma 0.2$, the ultimate strength σu, the relative elongation δ, and microhardness were determined (**Table 1**). Six types of samples were studied: TiNi in the initial state (as-received),

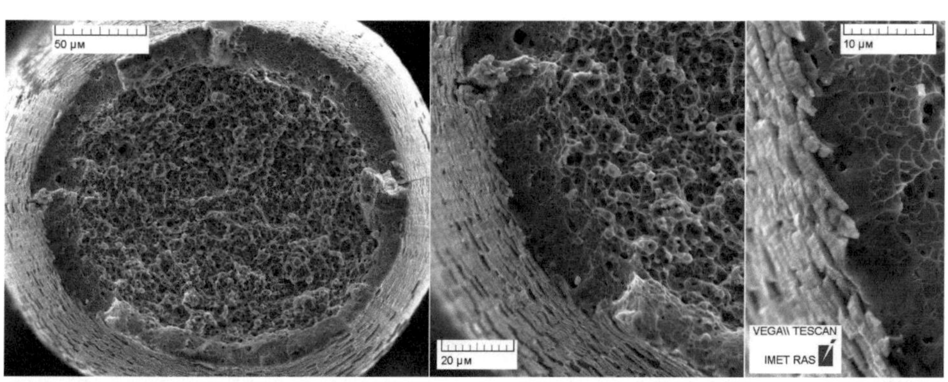

Figure 14.
Surface layer in a Ta-nitinol sample on a wire substrate after 80 min of sputtering with rotation at 865 mA, 400 V and distance of 150 mm.

Sample	σy, MПa	σu, MПa	δ, %	Microhardness, HV
TiNi	547 ± 5	1585 ± 7	47 ± 1	332 ± 3
TiNi after mechanical surface treatment and annealing	641 ± 6	1815 ± 9	54 ± 1	399 ± 3
TiNi after PIE	—	—	—	310 ± 6
Ta@TiNi, 10 min of sputtering	—	—	—	330 ± 4
Ta@TiNi, 30 min of sputtering	652 ± 7	1884 ± 8	55 ± 1	418 ± 4
Ti@TiNi, 30 min of sputtering	648 ± 6	1879 ± 8	55 ± 1	391.4 ± 5

Table 1.
Mechanical properties of composites based on nanostructured nitinol with surface layers of tantalum and titanium.

TiNi after polishing and annealing, TiNi after PIE, and composites with Ta and Ti layers. Wires after polishing were annealed at 450°C for 15 min in air as this treatment is needed for end stabilization of the nitinol structure that caused SME and superelasticity and product shaping. Composites were produced at 865 mA, 400 V, a sputtering distance of 150 mm and the conventional sputtering time listed in the **Table 1**.

Results of mechanical stretching tests show positive influence of 1 μm thickness surface layers of Ta and Ti on static properties of a nanostructural alloy, which promotes increase of yield strength and tensile strength by 2–4%. Relative elongation of all samples was of 55%. During preliminary ion etching of the surface of the substrate, bombardment with argon ions is carried out, which facilitates the removal of the surface oxide and the riveted layer with residual surface stresses and defects. Apparently, this explains a slight decrease in the microhardness of the samples immediately after PIE. Two types of composites were studied with a tantalum surface layer obtained for 10 (the main phase β-Ta) and 30 min (the main phase α-Ta formed on the beta phase). The surface of the composite material is distinguished by large microhardness values in comparison with the samples after PIE, since the hardness of both β-Ta and α-Ta is higher than that of nitinol. A thicker surface layer corresponds to higher microhardness values. With respect to the nanostructured substrate, a Ta layer of the order of 1 μm thick, consisting of a mixture of beta and alpha phases, shows an increase in the microhardness by about 26%. In this case, the effect of mechanical surface treatment and annealing on the surface microhardness can be practically neglected in connection with the PIE being carried out. The surface of the titanium layer less significantly affects the microhardness of the nanostructured substrate, but still increases it by 18%.

An object for investigations of corrosion resistance was wires of nanostructural nitinol and composites based on it with tantalum or titanium surface layers. Six types of samples were studied: (1) TiNi in the initial state (as-received), (2) TiNi after annealing, (3) TiNi after polishing, (4) TiNi after polishing and annealing, (5) TiNi-4 with Ta surface layer (Ta-TiNi), and (6) TiNi-4 with Ti surface layer (Ti-TiNi). Composites were produced at 865 mA, 400 V, a sputtering distance of 150 mm, a sputtering time 80 min on the main surface with rotation and 30 min on end faces after PIE. The researched composite materials had layered structure "a surface layer from the deposited substance (thickness ~ 0.9 microns)—the transitional layer containing elements both of the surface layer and of a basis (thickness ~ 0.2 microns)—a basis."

The material was tested for corrosion resistance under static conditions by dipping into solutions with various acidities because pH in the human body changes from 1 to 9. Neutral 0.9 wt% sodium chloride solution, artificial plasma and saliva, and four standard buffer solutions to reproduce acidic and alkaline media at the given level, and prepared from corresponding standard trimetric substances (fixanals) made by Merck, were used and listed in **Table 2**. Wire samples with a weight by 32.6 mg (separately from each other) were placed into flasks with 100 mL of the selected solution and aged totally in a dark place for up to 730 days. Sampling from flaks for analysis was after a selected period (7, 14, 30, 60, 90, 180, 360, or 730 days). The initial buffer solutions were used as reference solutions. Analysis was carried out by an ULTIMA 2 sequential atomic emission spectrometry (HORIBA Jobin Yvon, Japan) for using atomic emission spectrometry (AES) with inductively coupled plasma (ICP) for direct simultaneous determination of titanium and nickel in solutions. After immersion, the surface morphology and layer-by-layer composition were also investigated.

In **Figures 15** and **16**, an release of metal ions in model media depending on holding time, material treatment, nature, and temperature of the environment is shown. There are no results about all samples in the alkaline environment, artificial plasma and saliva, and also about TiNi-3, TiNi-4, Ta-TiNi, and Ti-TiNi samples in solutions with acidity 3.56–6.31, since in these cases, dissolution of elements was zero or below a limit of detection for all the time of a research. So, all further results concern only a solution with a pH of 1.68.

In the remained cases, elements' concentration in solutions increases (**Figures 15** and **16**) over time, but leaching of elements in medium considerably slows down. It can be related to sequential processes of the destruction and renewal of the protective film (de- and repassivation) on defect areas [30, 31].

Medium temperature growth insignificantly increases concentration of elements in solution (varies depending on immersion time and metal nature), but at the same time, the gradual inhibition of material dissolution is also observed, and at different temperatures, it occurs almost at the same time (**Figure 15**). It allows to assume that after initial increase in corrosion due to temperature increase in the following, with the surface repassivation, the degree of the material dissolution practically does not depend on temperature.

In solution with Ti-TiNi (**Figure 15a**), titanium concentration is approximately twice more than nickel that is explained by chemical interaction of surface layer material with potassium tetraoxalate [31]. In case of composite material with a tantalum surface layer, Ta concentration was also considered (**Figure 15b**). Insignificant dissolution of material is also observed only in the most acidic environment (most likely on possible defective sites of a surface with an incomplete surface layer that requires separate studying), and concentration of tantalum is much less than of titanium, which, respectively, is less, than of nickel.

pH	Composition
1.68	Potassium tetraoxalate: $KH_3C_4O_8 \times 2H_2O$, 0.05 M
3.56	Acid potassium tartrate: $C_4H_5O_6K$, 0.025 M
4.01	Acid potassium phthalate: $C_8H_5O_4K$, 0.05 M
6.31	Sodium chloride: NaCl, 0.9 wt.%
9.18	Acid sodium tetraborate: $Na_2B_4O_7 \times 10H_2O$, 0.05 M
7.36	Artificial plasma: NaCl (92.3 mM), $NaHCO_3$ (26.3 mM), K_2HPO_4 (0.9 mM), KCl (2.7 mM), NaH_2PO_4 (0.22 mM), $CaCl_2$ (2.5 mM), $MgSO_4 \cdot 7H_2O$ (0.82 mM), Na_2SO_4 (1.48 mM), D-glucose $C_6H_{12}O_6$ (5.55 mM) [23, 30, 31]
7.55	Artificial saliva: NaCl (13.34 mM), $NaHCO_3$ (7.4 mM), K_2HPO_4 (4.4 mM), KCl (10 mM), NaH_2PO_4 (1.2 mM), $CaCl_2$ (1.4 mM), $MgSO_4 \cdot 7H_2O$ (0.7 mM), Na_2SO_4 (0.13 mM), Na_2S (0.021 mM), carbamide (1 g/l) [23, 30, 31]

Table 2.
The composition and acidity of modeling solutions used for immersion test.

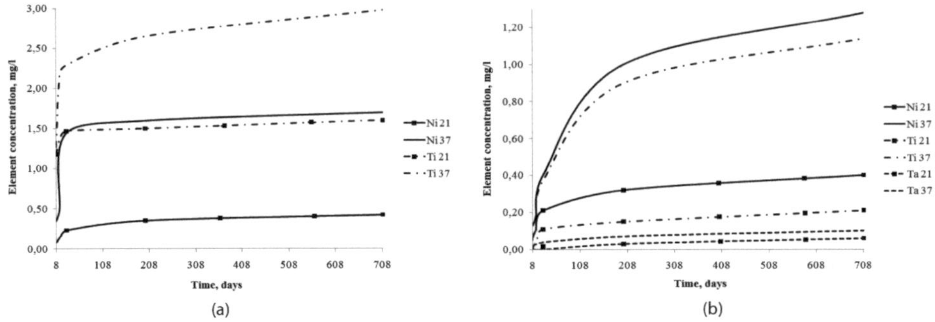

Figure 15.
Dependence of concentration of the elements dissolved from TiNi-Ti (a) and TiNi-Ta (b) composites in buffer solution with acidity 1.68 on immersion time of sampling and temperature of solution: the marked curves correspond to temperature of 21°C, curves without tags—37°C.

Depending on material treatment (**Figure 16**), ion release decreases in the following order: TiNi-2 > TiNi-1 > TiNi-4 > TiNi-3 > Ti-TiNi (if to look on nickel concentration) > Ta-TiNi. According to the literature, the thermal treatment at a temperature from 400 to 1000°C, which is required for stabilization of the mechanical properties, always results in a significant worsening of the corrosion resistance [31]. At the same time, the surface treatment, which facilitates the formation of the most perfect and homogeneous passive film, increases the corrosion resistance. Because of chemical interaction of Ti with acid media composite with its surface layer obviously less corrosion resistant than with Ta, but they are both more resistant than nitinol without a protective layer.

Composites Ta-TiNi and Ti-TiNi were also tested for biocompatibility.

The effect on the formation of H_2O_2 in the phosphate buffer (pH 6.8) on heating (37°C) for 200 min in the enhanced chemiluminescence system (luminol–p-iodophenol–peroxidase) [32] was studied. Sputtering of titanium and tantalum decreases the concentration of hydrogen peroxide formed by approximately 40 (6.5 ± 0.5 nM) and 60% (4.5 ± 0.3 nM), respectively, both close to the media concentration 3.2 ± 0.2 nM. By using a fluorescence probe specific to the OH radicals,

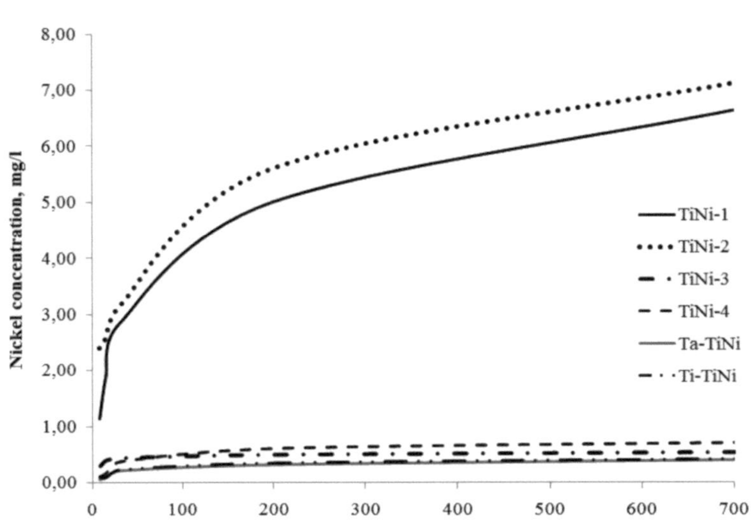

Figure 16.
Dependence of nickel concentration in buffer solution with acidity 1.68 at 21°C on immersion time and a sample type.

coumarin-3-carboxylic acid (Aldrich, USA) [33], it was found that all types of barrier coatings decrease the amounts of these radicals formed in a 20 mM phosphate buffer solution (pH 6.8) on heating (80°C) for 2 h. Titanium and tantalum coatings decreased the amount of the hydroxyl radicals by about 70 and 80% (30.9 ± 2.0 and 26.1 ± 1.3 nM), respectively, in comparison with NiTi plates (120.7 ± 4.9 nM). The test systems we used showed that the titanium or tantalum surface composite layers prevent the excessive generation of reactive oxygen species.

The biocompatibility was measured in vitro using standard test systems [34]. Then, the samples were examined under a DM 6000 fluorescence microscope (Leica, Germany). In the case of myofibroblasts from peripheral vessels, the percentages of vital cells for Ta-NiTi and Ti-NiTi were 95 ± 2 and 97 ± 2%, respectively. In the case of the human bone marrow mesenchymal stromal cells (MSC), the percentages of vital cells for Ta-NiTi and Ti-NiTi were 96 ± 3 and 96 ± 2%, respectively. Thus, none of the material surface samples used in the study had a short-term toxic effect on the cells that overgrew these surfaces de novo. The mitotic activity of the cells was assessed considering the mitotic index of the cells in the logarithmic growth phase (the third day after inoculation). The number of mitotic cells was determined by fluorescence microscopy using the vital staining with the Hoechst 33342 fluorescent dye (Sigma, USA). The MI value for the cells growing on the NiTi (reference) surface was 3.1% for the myofibroblast culture and 1.8% for the MSC culture. In the case of Ta-NiTi, the MI was 6.1% for the myofibroblasts and 4.3% for the bone marrow MSC. For the myofibroblasts and MSC cultured on Ti-NiTi, the mitotic indices were 5.8 and 4.7%, respectively. Morphological analysis of the myofibroblasts from peripheral vessels and bone marrow MSC on the surface of materials after 5 days of culturing were performed. Both myofibroblasts and MSC form a merged monolayer on the Ti-NiTi and Ta-NiTi surfaces.

3. Conclusions

Nano- and microdimensional surface layers of α- and β-Ta, Ti, Ag, and Cu on flat and wire NiTi, Cu, Ti, and SiO_2 substrates were created by vacuum magnetron sputtering aimed to investigate regularities of production of layered biomedical composite materials.

It was shown that the thickness and the structure of surface layers were affected by the sputtering distance, time, power, and the bias voltage at the substrate. The presence of the transition layer that contains both substrate and target elements and provides high adhesion of the surface layer to the substrate has been demonstrated. The morphology of a new surface repeats the substrate state regardless of the sputtering conditions.

With increase in deposition time, surface layer thickness does not linearly increase. Irrespective of summary sputtering time, the β phase is formed in the beginning, and at summary, surface layer thickness more than 0.6 μm on it α tantalum is deposited, while temperature remains below 150°C. The optimum bias voltage (500 V) for ion-atomic deposition was determined. It was demonstrated that an increase in power from 50 to 70% enhanced the thickness and uniformity of both the surface and the transition layers without their contamination.

A nonlinear increase in the thickness of the growing surface layers with decreasing sputtering distance under otherwise equal conditions was demonstrated. But the thickness of the transition layer and the dependence of the thickness change as a whole depend on the nature of the sputtered substance. It has been shown that at distances of 40–160 mm, insignificant deposition on the substrate side that is opposite to the sputtered flow is observed, with the thickness of formed layers also depending on the distance between the target and the substrate.

A slight corrosive dissolution was observed only in a medium with a pH of 1.56 for 2 years of a research. Dissolution in the other media is absent. Concentration of metals increases in solution over time, but the considerable slowdown of a metal ion release in solutions is observed over time. An increase in strength and plasticity in comparison with substrate was attained depending on the nature of the sputtered substance and substrate. Toxicity of samples has not been revealed.

Thus, the growth of a thin surface layer with high corrosion resistance and good biocompatibility by magnetron sputtering allows one to obtain a barrier to nitinol contact with physiological medium, which can withstand loads when nitinol exhibits superelasticity and an SME, with the formation of a transition layer, but no nitinol phase state changing.

Acknowledgements

The authors wish to thank Dyomin K Yu, Mikhailova AB, Gol'dberg MA, Kargin Yu F, and Gudkov SV for their help in sample analysis. The work was partially carried out under state assignment no. 007-00129-18-00 and was supported by the Russian President Program for Young Scientists (MK-4521.2018.8).

Chapter is partially based on the results of earlier published works [35–37], and authors have the permission to re-use it.

Conflict of interest

There is no conflict of interest to declare.

Author details

Elena O. Nasakina*, Mikhail A. Sevostyanov, Alexander S. Baikin, Sergey V. Konushkin, Konstantin V. Sergienko, Mikhail A. Kaplan, Ilya M. Fedyuk, Alexander V. Leonov and Alexey G. Kolmakov
A.A. Baikov Institute of Metallurgy and Material Science, Moscow, Russia

*Address all correspondence to: nacakina@mail.ru

References

[1] Stolin AM, Bazhin PM. Manufacture of multipurpose composite and ceramic materials in the combustion regime and high-temperature deformation (SHS extrusion). Theoretical Foundations of Chemical Engineering. 2014;**48**(6):751-763. DOI: 10.1134/S0040579514060104

[2] Bazhin PM, Stolin AM, Alymov MI. Preparation of nanostructured composite ceramic materials and products under conditions of a combination of combustion and high-temperature deformation (SHS extrusion). Nanotechnologies in Russia. 2014;**9**(11-12):583-600. DOI: 10.1134/S1995078014060020

[3] Maho A, Kanoufi F, Combellas C, Delhalle J, Mekhalif Z. Electrochemical investigation of nitinol/tantalum hybrid surfaces modified by alkylphosphonic self-assembled monolayers. Electrochimica Acta. 2014;**116**:78-88. DOI: 10.1016/j.electacta.2013.11.008

[4] Krivoshapkin PV, Mikhaylov VI, Krivoshapkina EF, Zaikovskii VI, Melgunov MS, Stalugin VV. Mesoporous Fe-alumina films prepared via sol–gel route. Microporous and Mesoporous Materials. 2015;**204**:276-281. DOI: 10.1016/j.micromeso.2014.12.001

[5] Kononova SV, Romashkova KA, Kruchinina EV, Gusarov VV, Potokin IL, Korytkova EN, Maslennikova TP. Polymer-inorganic nanocomposites based on aromatic polyamidoimides effective in the processes of liquids separation. Russian Journal of General Chemistry. 2010;**80**(6):1136-1142. DOI: 10.1134/S1070363210060162

[6] Nasakina EO, Baikin AS, Sevost'yanov MA, Kolmakov AG, Zabolotnyi VT, Solntsev KA. Properties of nanostructured titanium nickelide and composite based on it. Theoretical Foundations of Chemical Engineering. 2014;**48**(4):477-486. DOI: 10.1134/S0040579514040071

[7] Kuz'michev AI. Magnetronnye raspylitel'nye sistemy. Kn. 1. Vvedenie v fiziku i tekhniku magnetronnogo raspyleniya [Magnetron Scattering Systems. Book 1. Introduction into Physics and Technique of Magnetron Scattering]. Kiev: Avers; 2008. 244 p. (Russian)

[8] Bunshah RF. Deposition Technologies for Films and Coating. Park Ridge, New Jersey (USA): Noyes Publications; 1982

[9] Poate JM, Foti G, Jacobson DC. Surface Modification and Alloying by Laser, Ion and Electron Beams. New York: Plenum; 1983

[10] Dorranian D, Solati E, Hantezadeh M, Ghoranneviss M, Sari A. Effects of low temperature on the characteristics of tantalum thin films. Vacuum. 2011;**86**:51-55. DOI: 10.1016/j.vacuum.2011.04.012

[11] Bernoulli D, Müller U, Schwarzenberger M, Hauert R, Spolenak R. Magnetron sputter deposited tantalum and tantalum nitride thin films: An analysis of phase, hardness and composition. Thin Solid Films. 2013;**548**:157-161. DOI: 10.1016/j.tsf.2013.09.055

[12] Zhou YM, Xie Z, Ma YZ, Xia FJ, Feng SL. Growth and characterization of Ta/Ti bi-layer films on glass and Si (111) substrates by direct current magnetron sputtering. Applied Surface Science. 2012;**258**:7314-7321. DOI: 10.1016/j.apsusc.2012.03.176

[13] Navid AA, Chason E, Hodge AM. Evaluation of stress during and after sputter deposition of Cu and Ta films. Surface and Coating Technology.

2010;**205**:2355-2361. DOI: 10.1016/j.surfcoat.2010.09.020

[14] Myers S, Lin J, Martins Souza R, Sproul WD, Moore JJ. The β to α phase transition of tantalum coatings deposited by modulated pulsed power magnetron sputtering. Surface and Coating Technology. 2013;**214**:38-45. DOI: 10.1016/j.surfcoat.2012.10.061

[15] Cacucci A, Loffredo S, Potin V, Imhoff L, Martin N. Interdependence of structural and electrical properties in tantalum/tantalum oxide multilayers. Surface and Coating Technology. 2013;**227**:38-41. DOI: 10.1016/j.surfcoat.2012.10.064

[16] Navid AA, Hodge AM. Nanostructured alpha and beta tantalum formation—Relationship between plasma parameters and microstructure. Materials Science and Engineering A. 2012;**536**:49-56. DOI: 10.1016/j.msea.2011.12.017

[17] Navid AA, Hodge AM. Controllable residual stresses in sputtered nanostructured alpha-tantalum. Scripta Materialia. 2010;**63**:867-870. DOI: 10.1016/j.scriptamat.2010.06.037

[18] Zabolotnyi VT. Ionnoe peremeshivanie v tverdykh telakh [Ion Intermixing in Solids]. Moscow: MGIEM (TU); 1997 (Russian)

[19] Bose A, Hartmann M, Henkes H. A novel, self-expanding, nitinol stent in medically refractory intracranial atherosclerotic stenosis. The Wingspan Study, Stroke. 2007;**38**:1531-1537. DOI: 10.1161/STROKEAHA.106.477711

[20] Stoeckel D. Nitinol medical devices and implants. Minimally Invasive Therapy & Allied Technologies. 2000;**9**:81-88. DOI: 10.3109/13645700009063054

[21] Petrini L, Migliavacca F. Biomedical applications of shape memory alloys. Journal of Metallurgy. 2011;**2011**:1-15. DOI: 10.1155/2011/501483

[22] Zabolotnyi VT, Belousov OK, Palii NA, Goncharenko BA, Armaderova EA, Sevost'yanov MA. Materials science aspects of the production, treatment, and properties of titanium nickelide for application in endovascular surgery. Russian Metallurgy. 2011;**5**:437-448. DOI: 10.1134/S003602951105017X

[23] In: Ebrahim F, editor. Shape Memory Alloys—Fundamentals and Applications. Croatia, Rijeka: InTech d.o.o.; 2017. 134 p. Chapter 4. pp. 81-104. DOI: 10.5772/intechopen.69238

[24] Choi J, Bogdanski D, Köller M, Esenwein SA, Müller D, Muhr G, Epple M. Calcium phosphate coating of nickel–titanium shape-memory alloys. Coating procedure and adherence of leukocytes and platelets. Biomaterials. 2003;**24**:3689-3696. DOI: 10.1016/S0142-9612(03)00241-2

[25] Tan L, Dodd RA, Crone WC. Corrosion and wear-corrosion behavior of NiTi modified by plasma source ion implantation. Biomaterials. 2003;**24**:3931-3939. DOI: 10.1016/S0142-9612(03)00271-0

[26] Chen Y-H, Hsu C-C, He J-L. Antibacterial silver coating on poly (ethylene terephthalate) fabric by using high power impulse magnetron sputtering. Surface and Coating Technology. 2013;**232**:868-875. DOI: 10.1016/j.surfcoat.2013.06.115

[27] Cheng Y, Cai W, Li HT, Zheng YF. Surface modification of NiTi alloy with tantalum to improve its biocompatibility and radiopacity. Journal of Materials Science. 2006;**41**:4961-4964. DOI: 10.1007/s10853-006-0096-6

[28] Zhang M, Yang B, Chu J, Nieh TG. Hardness enhancement in nanocrystalline tantalum thin films. Scripta Materialia. 2006;**54**:1227-1230. DOI: 10.1016/j.scriptamat.2005.12.027

[29] Zhou YM, Xie Z, Xiao HN, Hu PF, He J. Effects of deposition parameters on tantalum films deposited by direct current magnetron sputtering. Journal of Vacuum Science and Technology A. 2009;**83**:286-291. DOI: 10.1116/1.304614

[30] Nasakina EO, Sevostyanov MA, Golberg MA, et al. Long term corrosion tests of nanostructural nitinol of (55.91 wt% Ni, 44.03 wt% Ti) composition under static conditions: Composition and structure before and after corrosion. Inorganic Materials: Applied Research. 2015;**6**(1):53-58. DOI: 10.1134/S2075113315010104

[31] Nasakina EO, Sevostyanov MA, Golberg MA, et al. Long term corrosion tests of nanostructural nitinol of (55.91 wt% Ni, 44.03 wt% Ti) composition under static conditions: ion release. Inorganic Materials: Applied Research. 2015;**6**(1):59-66. DOI: 10.1134/S2075113315010116

[32] Shtarkman IN, Gudkov SV, Chernikov AV, Bruskov VI. Effect of amino acids on X-ray-induced hydrogen peroxide and hydroxyl radical formation in water and 8-oxoguanine in DNA. Biochemistry (Moscow). 2008;**73**:470-478. DOI: 10.1134/S0006297908040135

[33] Bruskov VI, Karp OE, Garmash SA, et al. Free Radical Research. 2012;**46**:1280-1290. DOI: 10.3109/10715762.2012.709316

[34] Garmash SA, Smirnova VS, Karp OE, et al. Pro-oxidative, genotoxic and cytotoxic properties of uranyl ions. Journal of Environmental Radioactivity. 2014;**127**:163-170. DOI: 10.1016/j.jenvrad.2012.12.009

[35] Nasakina EO, Sevost'yanov MA, Mikhailova AB, Gol'dberg MA, Demin KY, Kolmakov AG, Zabolotnyi VT. Preparation of a nanostructured shape memory composite material for biomedical applications. Inorganic Materials. 2015;**51**(4):400-404

[36] Nasakina EO, Sevostyanov MA, Mikhaylova AB, Baikin AS, Sergienko KV, Leonov AV, Kolmakov AG. Formation of alpha and beta tantalum at the variation of magnetron sputtering conditions. IOP Conference Series: Materials Science and Engineering. 2016;**110**:012042

[37] Nasakina EO, Seregin AV, Baikin AS, Kaplan MA, Konushkin SV, Sergiyenko KV, et al. Formation of biocompatible surface layers depending on the sputtering distance. IOP Conference Series: Journal of Physics: Conference Series. 2017;**857**:012032

Chapter 2

Polyaniline/ZnO Nanocomposite: A Novel Adsorbent for the Removal of Cr(VI) from Aqueous Solution

Rais Ahmad

Abstract

In recent years with the rapid economic globalization, pollution by industries and agriculture has increased, which results in decrease in the quality of ground and surface water. Pollution by heavy metals has become a serious health issue worldwide due to their nonbiodegradable and persistent nature. Therefore, extensive research has been done to develop ecofriendly and effective methods for removal of heavy metals, such as chemical precipitation, ion exchange, electrodialysis, membrane filtration, and adsorption. Among these methods, adsorption is the most recognized technique for wastewater treatment due to high-removal efficiency and ease in operation without yielding harmful by-products. Recently, nanocomposites based on biopolymer-grafted synthetic adsorbent have been used in various industrial applications including wastewater treatment. Therefore, the present chapter will be devoted for the removal of toxic heavy metals from wastewater by using bionanocomposite.

Keywords: heavy metals, bionanocomposite, adsorption, desorption

1. Introduction

In today's era, the industrialization, agriculture, and domestic activities have led to a large amount of wastewater having toxic elements such as lead, mercury, cadmium, copper, arsenic, chromium, etc., and dyes which has adversely affected human lives, animal lives, and environment. Among various heavy metals, Cr(VI) is one of the most important and toxic heavy metals due to its vast applications in industries [1, 2]. In aqueous medium, two forms of chromium exist, i.e., Cr(III) and Cr(VI), and the toxicity and reactivity of both the forms mainly depend on oxidation state of the chromium [3]. In trace amounts, Cr(III) is an essential nutrient for humans and to mammals for their maintenance of normal glucose tolerance factor, lipid, and protein metabolism [4]. On the other hand, Cr(VI) is very toxic to human as well as marine life and poses serious health problems such as liver damage and pulmonary congestion and regarded as carcinogenic also [5–7].

Various conventional methods have been developed for the adsorption of Cr(VI) in wastewater, including electrochemical precipitation, ion exchange, membrane ultrafiltration, reverse osmosis, reduction, and adsorption [8–10]. Adsorption has been widely used for the removal of chromium from contaminated groundwater out of these methods [11] due to its low initial cost and ease of operation and high efficiency to remove toxic heavy metals. This technique can be harnessed at large scale for the treatment of polluted water as it can handle fairly large flow rates, producing a high quality of water without producing notorious sludge and residual contaminants [12–14].

Conducting polymer with metal oxide has nowadays emerged as an attractive alternative for the sequestration of wastewater, as it mainly provide larger surface area for the adsorption, interfacial adhesion between the surface of nanocomposite and metal ions, it is easily tractable and cost effective. Due to large amount of amine and imine functional groups PANI has strong affinity with metal ions and can remove heavy metal contaminants from aqueous solutions effectively [15, 16]. The stability of polymer matrix can be enhanced by addition of fillers such as natural clays, metal oxide nanoparticles, etc. [17]. Zinc oxide is known to be a potential adsorbent of Cr(VI) at high temperatures as well as at low temperatures [18]. So, the incorporation of ZnO nanoparticles in the PANI matrix not only provides thermal and mechanical strength, but also improves its adsorption properties. This improvement was attributed to the increase in the number and high dispersion of active terminal OH groups of ZnO nanoparticles with amine groups of PANI.

In the present chapter, polyaniline zinc oxide nanocomposite was synthesized by oxidative free radical polymerization of aniline monomer in presence of zinc oxide nanoparticles. The material was characterized by various analytical techniques such as FT-IR, XTD, TGA-DTG, SEM, and TEM. The nanocomposite material was further explored for the removal of Cr(VI) from aqueous solution. The effects of various adsorption parameters viz. agitation time, solution pH, adsorbent dose, initial metal ion concentration, and temperature was observed and optimized by preliminary experiments.

2. Materials and methods

2.1 Chemicals

Zinc acetate and ammonium persulfate (APS) were procured from Sigma Aldrich, India. Sodium hydroxide and potassium dichromate were procured from Merck, India. Aniline monomer was procured from Fisher Scientific, India and was distilled before use. About 1000 mg L^{-1} stock solutions of were prepared by dissolving appropriate amount of metal salts in double distilled water.

2.2 Synthesis of zinc oxide nanoparticles

ZnO nanoparticles were synthesized by direct precipitation method using zinc acetate and KOH as precursors [19]. The aqueous solution (0.2 M) of zinc acetate ($Zn(CH_3COO)_2\ 2H_2O$) and the solution (0.4 M) of KOH were prepared with deionized water, respectively. The KOH solution was slowly added into zinc nitrate solution at room temperature under vigorous stirring, which resulted in the formation of a white suspension. The white product was centrifuged at 5000 rpm for 20 min and washed three times with distilled water, and washed with absolute alcohol at last. The obtained product was calcined at 500°C in air atmosphere for 3 h.

2.3 Synthesis of polyaniline/ZnO nanocomposite

The material was synthesized by methods already reported elsewhere [20]. The detailed procedure is as follows: 1 g of above synthesized ZnO nanoparticles were taken in 100 ml of 0.1 M HCl solution and sonicated for 1 h at 30°C. Now, 10 ml of distilled aniline monomer taken in 400 ml of 0.1 M HCl solution was poured in the above colloidal solution of ZnO. The mixture was allowed to stir for 2 h for complete mixing at 5°C. The reaction was proceeded with the addition of 15 g of APS. The reaction was allowed to stand for 12 h at 5°C and product obtained was washed with 0.1 M HCl solution five to six times in order to remove excess amount of unreacted aniline.

2.4 Instrumentation

The FTIR spectra of the adsorbent materials were recorded with a Perkin Elmer 1800 model IR spectrophotometer operating at frequency range from 400 to 4000 cm^{-1} using KBr pallets. X-ray diffraction (XRD) patterns of the samples were obtained using Siemens D 5005 X-ray unit Cu Kα (λ = 1:5406 Å) radiation, generated at a voltage of 40 kV and a current of 40 mA was used as the X-ray source. Scanning electron microscopy and electron diffraction scattering (SEM/EDS) analysis were done using GSM 6510LV scanning electron microscope. The particle size and structure of the synthesized nanocomposite were observed by using JEM 2100 transmission electron microscope (TEM). The thermal stability was determined by derivative thermal analysis (DTG, Perkin Elmer Pyris 6) and DTA (Perkin Elmer model, STA 6000). The thermograms were recorded for 20 mg of powder sample at a heating rate of 10°C min^{-1} in the temperature range of 30–800°C under nitrogen atmosphere. The concentration of metal ions in the solution was measured by atomic absorption spectrophotometer (AAS) model GBC-902. Elico Li 120 pH meter was used to adjust the pH of the solutions.

2.5 Adsorption experiments

The adsorption experiments were performed using batch equilibrium technique in aqueous solutions in the temperature range of 30–50°C at pH 2. For this, 0.04 g of adsorbents was added to 20 ml of Cr(VI) solution of various concentrations (from 10, 50, 100 mg L^{-1}) and shaken in a thermostatic water-bath shaker operated at 120 rpm. After equilibrium was attained, the adsorbent was removed and the supernatant was collected after attaining equilibrium. The concentrations of Cr(VI) in supernatant were measured using atomic absorption spectrophotometer (AAS). The kinetic experiments were performed at three different Cr(VI) ion concentrations mainly at 10, 50, and 100 mg L^{-1} at pH 2 and 3. The effect of time on the adsorption of Cr(VI) ions on PAZO nanocomposite was studied at 10–300 min and the equilibrium was reached at 120 min. The effect of adsorbent dose and initial metal ion concentration was also studied. The amount of metal ions adsorbed onto PAZO was calculated by a mass balance relationship:

$$q_e = \frac{(C_o - C_e)V}{W} \tag{1}$$

where q_e is the amount of metal ion adsorbed per unit weight of the adsorbent (mg g^{-1}); C_o and C_e are the concentrations of the metal ion in the initial solution (mg L^{-1}) and after adsorption, respectively; V is the volume of the adsorption medium (L); and W is the amount of the adsorbent (g).

3. Results and discussion

3.1 Characterization of PAZO

3.1.1 XRD analysis

XRD analysis of bulk composite was done and it was seen that the characteristic peak of ZnO nanoparticles were obtained at 2θ values of 8.03, 11.87, 31.71, 34.38, 38.18, 47.47, 58.51, 62.77, and 68.98° given in **Figure 1** which corresponds to Miller indices (100), (002), (101), (102), (110), (103), and (201), respectively. The crystalline size calculated using Scherer formula was found to be 31 nm which is also confirmed by TEM micrograph. The PANI peak diffracted at an angle of 25.72°, respectively, in the XRD pattern showing low crystallinity for conductive polymers due to the repetition of benzenoid and quinoid rings in PANI chains. Peaks of PANI-ZnO composites shift slightly higher values of 2θ. It can be seen that the XRD patterns of nanocomposites represent the peaks from ZnO and PANI. This is because of the presence of ZnO nanoparticles which is equal to 5% and it has significant effect on diffraction pattern of PANI. According to **Figure 1**, two distinct sharp peaks at 2θ = 19.311 and 25.721 with planes of (010) and (200), respectively, are shifted negligibly but their intensity increases by the reinforcement of the ZnO nanoparticles in PANI matrix. Additionally, one peak at 2θ = 23.21 with plane of (102) appears which is related to PANI-CSA and its intensity is increased by adding ZnO nanoparticles. XRD results confirm the effect of ZnO nanoparticles in PANI-ZnO nanocomposites. **Figure 1** shows that intensity of the peaks was increased by incorporation of 5% ZnO nanoparticles, which means that there is an interaction of ZnO nanoparticles and PANI by formation of hydrogen bonding between H–N and oxygen of ZnO [21].

3.1.2 Thermal analysis

Thermal stability of PANI and PAZO was analysed by TGA and the thermograms are given in **Figure 2**. The TGA thermogram of pure PANI shows three-step

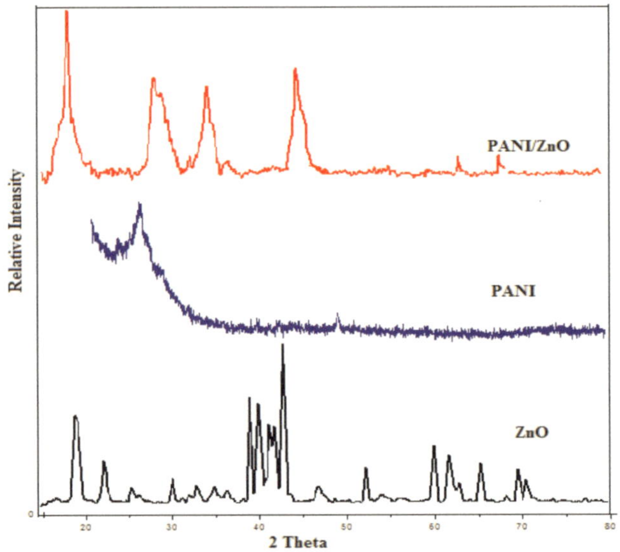

Figure 1.
XRD spectra of ZnO, PANI and PAZO.

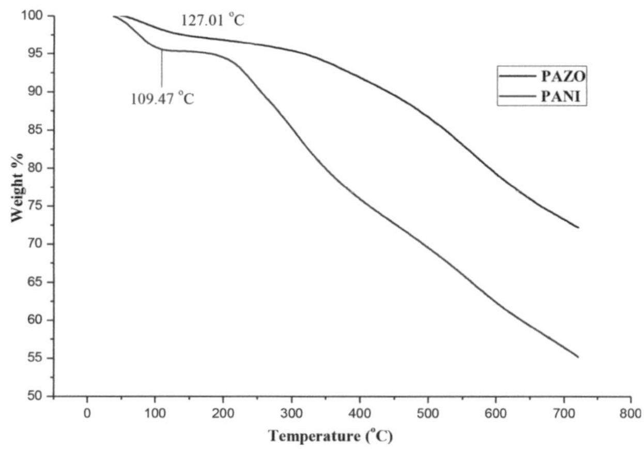

Figure 2.
TGA thermograms of PANI and PAZO.

degradation of weight in the range of 30–800°C. The first weight loss of at around 109.47°C is due to evaporation of water. The second stage of weight loss starting at around 140°C up to 370°C almost 60% substance weight loss which represents the degradation of low molecular weight polymers and almost 45% weight loss for PANI was observed at 700°C [22]. From 370°C onward, degradation of PANI chains takes place up to 800°C, in which almost 90% mass loss is observed. The PAZO also shows same stages of weight loss with little bit of higher thermal stability as compare to pure PANI due to incorporation of ZnO in PANI matrix and a weight loss of 28% was observed at 800°C.

3.1.3 Surface analysis (SEM with EDX)

Morphologies of the PANI-ZnO nanocomposite with its EDX image before and after adsorption of Cr(VI) are shown in **Figure 3**. Nanocomposite reveals flaky fibrous structure shaped structure. These nanocomposites should give the opportunity to obtain improved capacitance due to surface effects. The size of the flakes and fibers decreased due to adsorption of Cr(VI). The SEM images help us to draw a conclusion that the doping of ZnO nano-rods has a strong effect on the morphology of PANI, since PANI has various structures such as granules, nanofibers, nanotubes, nanospheres, microspheres, and flakes.

3.1.4 Transmission electron microscope (TEM) analysis

Due to the low magnification in SEM micrographs, it is difficult to observe ZnO nanoparticles in the nanocomposite matrix; thus, an appropriate way for observing them in polymer matrix is by the use of TEM. According to the TEM micrographs in **Figure 4**, PANI and ZnO nanoparticles have formed a nanocomposite in which the nanoparticles are embedded in the polymer matrix. It is obvious that ZnO nanoparticles were uniformly coated by PANI. The average size of ZnO nanoparticles was observed as 31.2 nm.

3.1.5 FTIR analysis

PANI and PAZO nanocomposite were characterized by using the FTIR technique. **Figure 5** shows the FTIR pattern of ZnO nanoparticles, PANI and PAZO nanocomposite. The characteristic absorption bands of PANI are 515.71 cm^{-1}

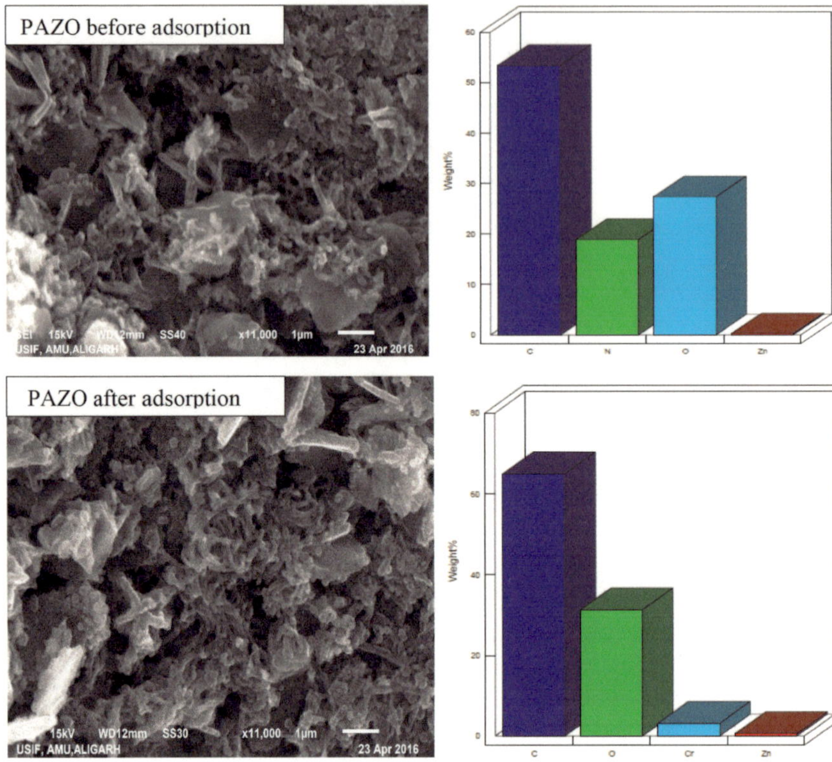

Figure 3.
SEM images of PAZO before and after adsorption of Cr(VI) with its EDX image.

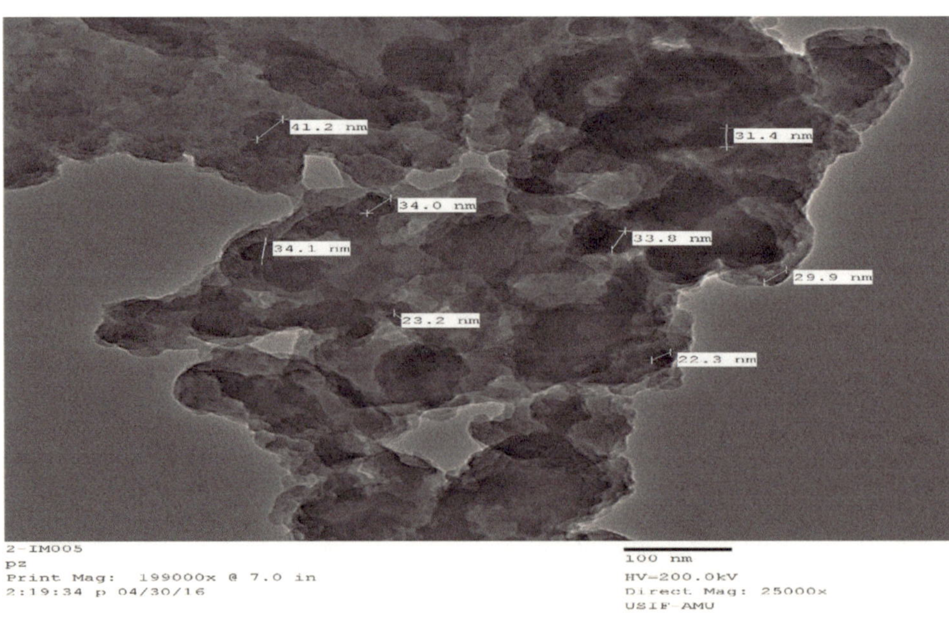

Figure 4.
TEM image of PAZO.

(C–N–C bonding mode of aromatic ring); 592.85 and 700.84 cm^{-1} (C–C, C–H bonding mode of aromatic ring); 831.98 cm^{-1} (C–H out of plane bonding in benzenoid ring); 1040.26, 1302.53, and 1503.09 cm^{-1} (C–N stretching of benzenoid ring), and 1572.52 cm^{-1} (C–N stretching of quinoid ring). The PAZO nanocomposite shows the same characteristic peaks. However, there is an evidence of peak displacement when ZnO nanoparticles are added to the PANI. These shifts include 1572.51–1587.94, 1503.09–1510.81, 1155.97–1148.25, 1040.26–1047.97, 831.98–862.84, and 592.85–600.23 cm^{-1}. Furthermore, in PAZO nanocomposite, a broad peak appeared in 3470 cm^{-1}, which can be associated to the interaction between ZnO nanoparticles and PANI by formation of hydrogen bonding between H–N and oxygen of ZnO. So, the peak displacement that was observed in FTIR spectra may be ascribed to the formation of hydrogen bonding between ZnO and the N–H group of PANI on the surface of the ZnO nanoparticles.

3.2 Adsorption behavior of PAZO toward Cr(VI)

3.2.1 Effect of pH

One the most important parameter that directly affects the adsorption of Cr(VI) is pH of the solution. The effect of the initial solution pH on the removal of Cr(VI) was studied with 0.04 g of PAZO nanocomposite, 20 ml of 10, 50, and 100 mg L^{-1} Cr(VI) solution with different pH in the range 2–7 at 30°C. The effect of pH on

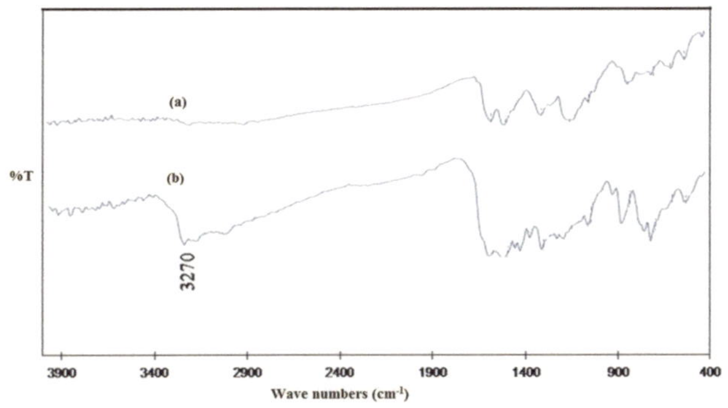

Figure 5.
FTIR spectra of (a) PANI and (b) PAZO.

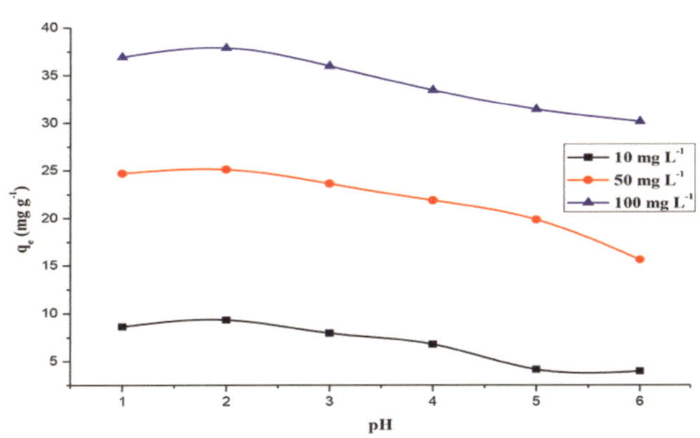

Figure 6.
Effect of pH on adsorption of Cr(VI) on PAZO at 10, 50, and 100 mg L^{-1} initial metal ion concentrations.

sorption of Cr(VI) has been shown in **Figure 6**. It was found that the maximum adsorption capacity for Cr(VI) was in the pH value of 2 and as the pH value increases the adsorption capacity decreases. Various forms of Cr(VI) in water such as $HCrO_4^-$ in acidic medium, CrO_4^{2-} in neutral and basic medium are predominant factor for the adsorption of Cr(VI) onto PAZO nanocomposite.

At pH < 4, due to presence of excess of H^+ ions in the solution, the adsorbent surface becomes positively charged due to protonation and $HCrO_4^-$ form of Cr(VI) ions are dominant at lower pH [23], so strong electrostatic attraction between positively charged adsorbent surface and negatively charged $HCrO_4^-$ ions led to higher removal efficiency. However, as the pH increases, deprotonation of surface of the adsorbent was observed due to decrease in number of H^+ ions. So, lower adsorption capacity results due to less interaction between Cr(VI) ions and adsorbent surface at higher pH value. The point of zero charge of adsorbent surface is found to be 5.5.

3.2.2 Effect of contact time and initial metal ion concentration

The effect of retention time on removal efficiency of Cr(VI) was carried out by varying the contact time in the range of 10–300 min at three different metal ion concentrations 10, 50, and 100 mg L^{-1} at pH 2 at adsorbent dose of 0.04 g. The effect of contact time on PAZO for Cr(VI) removal is depicted in **Figure 7** indicating an initial increase in adsorption capacity with increase in time and attaining the equilibrium time at 120 min after that little change in adsorption capacity for Cr (VI) is seen which indicates that the system has already achieved equilibrium. No change in adsorption capacity after equilibrium reveals that the adsorption sites are completely occupied by metal ion. So, the equilibrium time 120 min was chosen as optimum time in subsequent experiments.

The initial metal ion concentration provides an important driving force to overcome all mass transfer resistance of metal ions between the aqueous and solid phases [24]. Three different concentrations 10, 50, and 100 mg L^{-1} of the metal ions were chosen to see the effect of initial metal ion concentration on adsorption capacity of PAZO. **Figure 7** also shows the effect of initial metal ion concentration in which by increasing Cr(VI) ions concentration the adsorption capacity also increases. The maximum adsorption capacity at 10, 50, and 100 mg L^{-1} was found to be 9.31, 25.11, and 31.89 mg g^{-1}, respectively, which might be due to the fact that increasing metal ion concentration increased the number of collision between the adsorbent and metal ion species, this leads to an increased metal ion uploading [25].

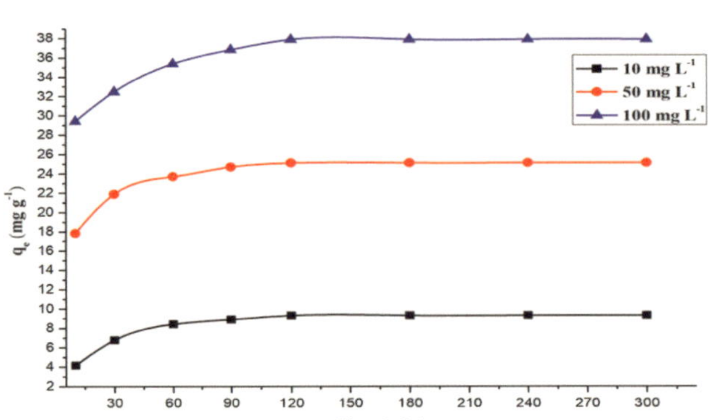

Figure 7.
Effect of contact time on adsorption of Cr(VI) on PAZO at 10, 50, and 100 mg L^{-1} initial metal ion concentrations.

3.2.3 Effect of adsorbent dose and temperature

The adsorption of metal ions in the solution is greatly affected by the dose of adsorbent used. A range of adsorbent dose from 0.01 to 0.07 g was used with 20 ml of 10, 50, and 100 mg L^{-1} of metal ion solution for 120 min to investigate the effect of dose on removal of Cr(VI), and the results are shown in **Figure 8**. It was found that the adsorption efficiency for Cr(VI) ion increases as the amount of adsorbent increase up to 0.04 g, but on further increasing the adsorbent dose, the adsorption capacity decreases. This trend can be explained as the adsorbent dose increases, the number of adsorbent particles also increases facilitating more active sites for adsorption but on further increase in adsorbent dose adsorption capacity decreases due to partial aggregation of adsorbent particles. So, 0.04 g was taken as optimum adsorbent dose for all the experiments.

The effect of temperature on adsorption of Cr(VI) was observed in the temperature range of 30–50°C (plot not given). It was observed that on increasing the temperature, the adsorption capacity also increases due to increase in diffusion rate of metal ions across the external boundary layer and within the pores of PAZO nanocomposite. Furthermore, at high temperature, the energy of the system also facilitated the binding of Cr(VI) on the surface of PAZO indicating the adsorption of Cr(VI) is controlled endothermic process. So, 50°C was selected as the optimum temperature for all the adsorption experiments.

3.3 Adsorption isotherms

The adsorption isotherms describe the effect of metal ion concentrations on the amount of metal ion adsorbed on the adsorbent surface leading to find the best equilibrium position in the adsorption process. In the present study, Langmuir [26], Freundlich [27] Dubinin-Radushkevich [28], and Temkin [29] models have been applied to the experimental data.

3.3.1 Langmuir isotherm

The Langmuir isotherm is used to describe the equilibrium between the surface of solid and the solution as a reversible chemical equilibrium. Langmuir isotherm model is valid for adsorption onto a surface containing a finite number of identical sites. The Langmuir treatment is based on the assumption that a maximum

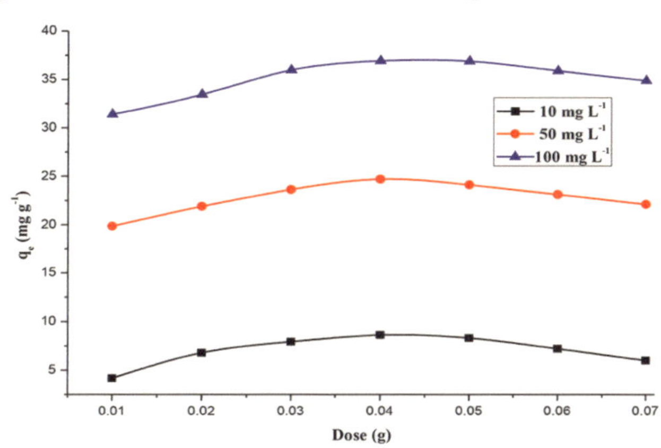

Figure 8.
Effect of adsorbent dose on adsorption of Cr(VI) on PAZO at 10, 50, and 100 mg L^{-1} initial metal ion concentrations.

adsorption corresponds to a saturated monolayer of solute molecules on the adsorbent surface which is represented as follows:

$$\frac{C_e}{q_e} = \frac{1}{q_m K_L} + \frac{C_e}{q_m} \tag{2}$$

where q_m is maximum monolayer adsorption capacity of the adsorbent (mg g^{-1}) and K_L is the Langmuir constant (L mg^{-1}) related to the adsorption free energy. The q_m and K_L values are reported in **Table 1**. The essential feature of the Langmuir adsorption can be expressed by means of R_L, a dimensionless constant referred to as separation factor for predicting whether an adsorption system is favorable or unfavorable. R_L is calculated using equation

$$R_L = \frac{1}{1 + K_L C_o} \tag{3}$$

where C_0 is the initial metal ion concentration (mg L^{-1}). The parameter R_L indicates the favorability of adsorption as follows:

- $R_L > 1$, unfavorable adsorption
- $0 < R_L > 1$, favorable adsorption
- $R_L = 0$, irreversible adsorption
- $R_L = 1$, linear adsorption

The linearized plot (**Figure 9**) of C_e/q_e versus C_e are obtained for Cr(VI). The Langmuir constants q_m and K_L can be determined from the slope and intercept of the linear line, respectively. As can be seen from **Table 1**, the maximum monolayer

Model	Parameters	Cr(VI)		
		30°C	40°C	50°C
Langmuir	qm (mg g^{-1})	120.92	134.22	139.47
	K_L (L mg^{-1})	0.06	0.08	0.12
	R^2	0.99	0.99	0.99
Freundlich	n	1.18	1.22	1.35
	K_L (mg g^{-1})	5.69	6.29	6.62
	R^2	0.98	0.98	0.98
D-R	q_m (mg g^{-1})	59.05	62.18	68.70
	k_{D-R} (mol^2 KJ^{-2})	$1.85 \times e^{-6}$	$1.47 \times e^{-6}$	$1.25 \times e^{-}$
	E (KJ mol^{-1})	10.48	11.71	12.65
	R^2	0.97	0.98	0.97
Temkin	A (L mg^{-1})	0.55	0.60	0.64
	b (J mol^{-1})	63.10	62.97	56.09
	R^2	0.98	0.98	0.98

Table 1.
Isotherm parameters for Cr(VI) removal by PAZO at 30, 40, and 50°C.

adsorption capacities (q_m) calculated by Langmuir model were found to be 120.92, 134.22, and 139.47 mg g^{-1} at 30, 40, and 50°C, respectively. The regression coefficient values calculated are 0.99, 0.99, and 0.99, respectively, which suggest that the Langmuir model is best fitted to the experimental data at all temperature ranges.

3.3.2 Freundlich isotherm

The Freundlich isotherm model is the earliest empirical equation based on the adsorption on reversible heterogeneous surfaces. The mathematical expression of the model is given as follows:

$$\ln q_e = \frac{1}{n} \ln C_e + \ln K_F \tag{4}$$

where K_F (mg g^{-1}) is approximately an indicator of the adsorption capacity, and $1/n$ is the adsorption intensity and an indicator for the favorability of adsorption. The values of $n > 1$ represent favorable adsorption condition. The linear plot (**Figure 10**) of ln q_e versus ln C_e gives slope of value $1/n$ and an intercept ln K_F. When C_e equals unity, ln K_F is equal to ln q_e. In the other case, when $1/n = 1$, the K_F value depends on the units in which q_e and C_e are expressed. A favorable adsorption tends to give Freundlich constant n value between 1 and 10. Larger value of n

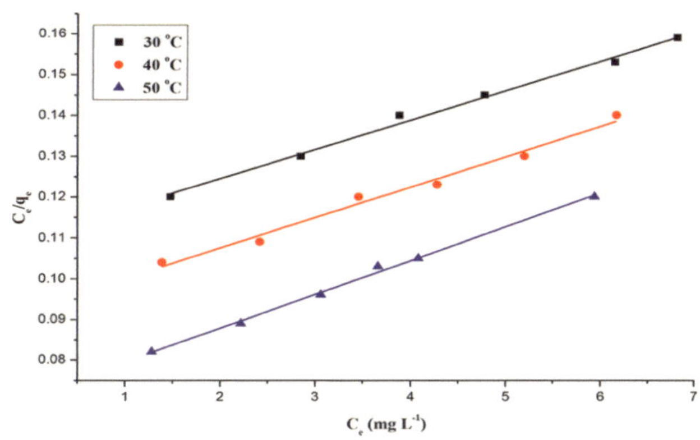

Figure 9.
Langmuir adsorption isotherm for Cr(VI) on PAZO at 30, 40, and 50°C (dose = 0.04 g and pH = 2).

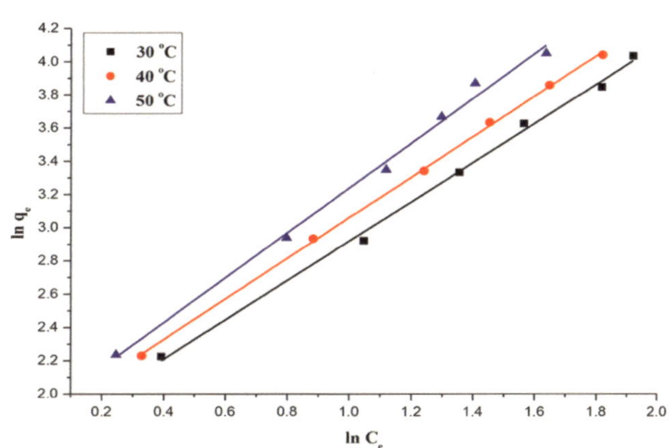

Figure 10.
Freundlich adsorption isotherm for Cr(VI) on PAZO at 30, 40, and 50°C (dose = 0.04 g and pH = 2).

(smaller value of $1/n$) implies strong interaction between sorbent and metal ions, while $1/n$ equal to 1 indicates linear adsorption leading to identical adsorption energies for all the sites. It can be observed from **Table 1** that for all the temperature ranges, the values of n is >1 and hence favorable adsorption.

3.3.3 Dubinin-Radushkevish (D-R) isotherm

The Dubinin-Radushkevish (D-R) isotherm can be used to describe adsorption on both homogeneous and heterogeneous surfaces. This isotherm can be described by the following equation:

$$\ln q_e = \ln q_m - k_{D-R}\varepsilon^2 \quad (5)$$

where q_m is the D-R monolayer capacity (mg g^{-1}) obtained by a plot between ln q_e and ε^2 (**Figure 11**), $k_{D\text{-}R}$ is a constant related to the adsorption energy, and $\boldsymbol{\varepsilon}$ is Polanyi potential which is related to the equilibrium concentration as follows:

$$\varepsilon = RT\left(1 + \frac{1}{C_e}\right) \quad (6)$$

where R is the gas constant (8.314 $Jmol^{-1}$ K^{-1}) and T is the absolute temperature (K). The constant k gives the mean free energy E of adsorption per molecule of the adsorbate when it is transferred to the surface of the solid from infinity in the solution and can be computed using the relationship:

$$E = \frac{1}{\sqrt{2k_{D-R}}} \quad (7)$$

The magnitude of E is useful for estimating the mechanism of the adsorption reaction. From **Table 1**, the maximum adsorption capacities for Cr(VI) calculated by D-R model at 30, 40, and 50°C was found to be 59.05, 62.18, and 68.70 mg g^{-1}, respectively. The mean free energy per molecule (E) was estimated to be 10.48, 11.71, and 12.65 KJ mol^{-1} at 30, 40, and 50°C, respectively, which confirms that the adsorption reaction follows chemisorption process. The values of D-R constant $k_{D\text{-}R}$ were found to be 1.85×10^{-6}, 1.47×10^{-6}, and 1.25×10^{-6} for Cr(VI) at 30, 40, and 50°C, respectively.

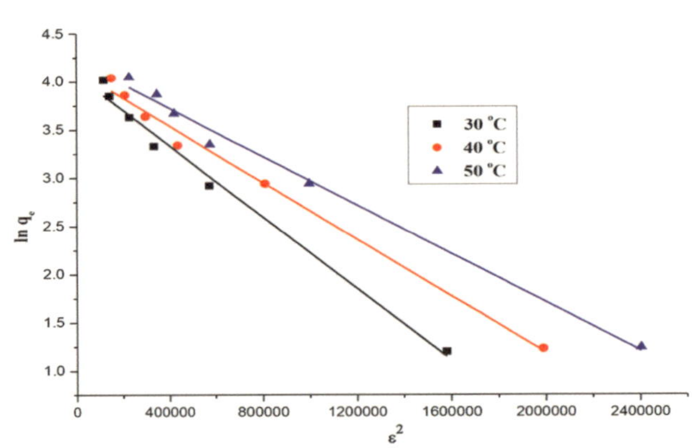

Figure 11.
D-R adsorption isotherm for Cr(VI) on PAZO at 30, 40, and 50°C (dose = 0.04 g and pH = 2).

3.3.4 Temkin isotherm

Temkin isotherm considers the effects of the heat of adsorption that decreases linearly with coverage of the adsorbate and adsorbent interactions. The Temkin isotherm has been used in the following forms:

$$q_e = B\ln A + B\ln C_e \tag{8}$$

$$B = \frac{RT}{b} \tag{9}$$

where R is gas constant 8.314 J mol^{-1} K^{-1}. T is absolute temperature (K), b is the Temkin constant related to the heat of adsorption (J mol^{-1}), and A is the equilibrium binding constant corresponding to the maximum binding energy (L g^{-1}). The liner plots (**Figure 12**) of q_e versus ln C_e enable to determine the constant A and B. The Temkin constant given in **Table 1** clearly suggests that the adsorption involves chemisorption and physisorption of the Cr(VI) rather than an ion exchange mechanism. The Temkin isotherm gives a satisfactory linear fit data with all the metal ions. The value of binding constant A given in **Table 1** as 0.55, 0.60, and 0.64 L mg^{-1} also support the high affinity of Cr(VI) toward adsorbent surface at high temperature.

3.4 Adsorption kinetics

Kinetics of the adsorption process is a vital parameter which provides essential information on the solute uptake rate and the reaction pathways. To determine the rate-determining step during the adsorption process, the kinetic data of heavy metals onto PAZO were simulated with the pseudo-first-order and pseudo-second-order, Elovich, and Webber-Morris intra-particle diffusion models.

3.4.1 Lagergren pseudo-first-order model

The pseudo-first-order kinetics model is based on the assumption that adsorption was controlled by diffusion steps [30] and the rate of adsorption is in direct proportion to the difference value of equilibrium adsorption capacity and the adsorption capacity at any time t. The linear equation for this model can be expressed by the following equation:

$$\log\left(q_e - q_t\right) = \log q_e - \frac{k_1}{2.303}t \tag{10}$$

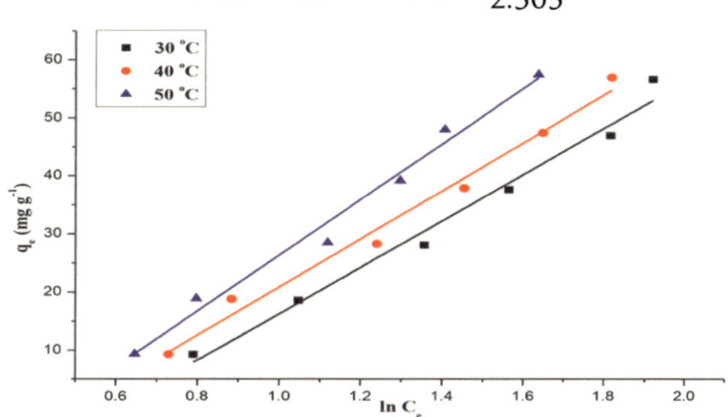

Figure 12.
Temkin adsorption isotherm for Cr(VI) on PAZO at 30, 40, and 50°C (dose = 0.04 g and pH = 2).

where k_1 is the pseudo-first-order rate constant (min^{-1}), q_e is the amount of heavy metal ions adsorbed at equilibrium ($mg\ g^{-1}$), and q_t is the amount of the adsorption at any time t ($mg\ g^{-1}$). Such an equation should yield a straight line, as given in **Figure 13**, with intercept equal to log q_e and slope equal to (k_1/2.303).

3.4.2 Pseudo-second order

The linear equation for pseudo-second-order kinetics [31] is given by the following equation:

$$\frac{t}{q_t} = \frac{1}{k_2 qe^2} + \frac{t}{q_e} \tag{11}$$

where k_2 is the pseudo-second-order rate constant ($g\ mg^{-1}\ min^{-1}$), q_e is the amount of heavy metal ions adsorbed at equilibrium ($mg\ g^{-1}$), and q_t is the amount of the adsorption at any time t ($mg\ g^{-1}$). The linear plot of pseudo-second-order model is given in **Figure 14**, from which constant k_2 and q_e can be calculated.

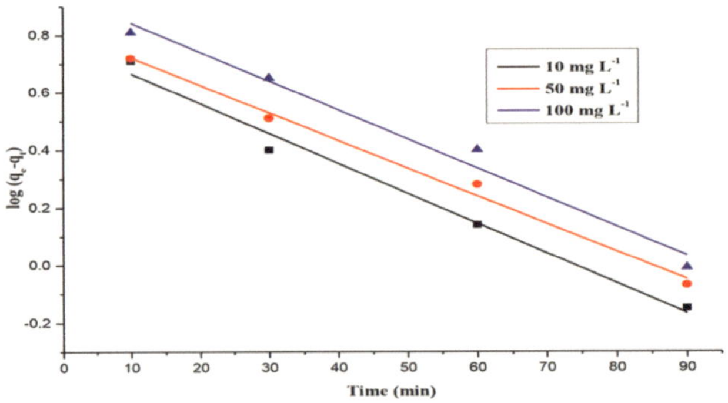

Figure 13.
Pseudo-first-order model for Cr(VI) on PAZO at 10, 50, and 100 mg L^{-1} metal ion concentration (dose = 0.04 g and pH = 2).

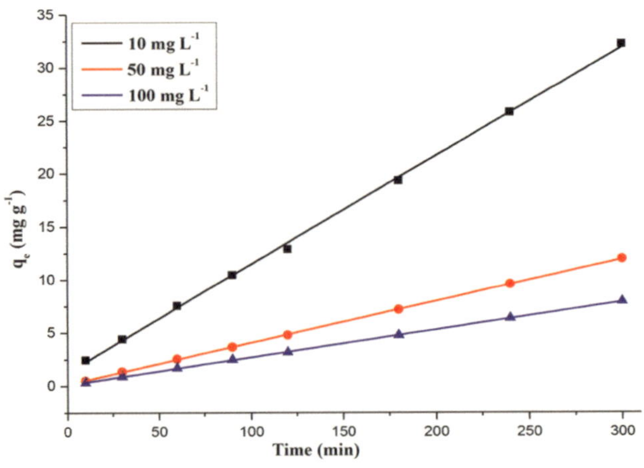

Figure 14.
Pseudo-second-order model for Cr(VI) on PAZO at 10, 50, and 100 mg L^{-1} metal ion concentration (dose = 0.04 g and pH = 2).

3.4.3 Elovich model

If the process is a chemisorption on highly heterogeneous sorbents, the sorption kinetics could be interpreted by Elovich equation [32] as follows:

$$q_t = \frac{1}{\beta}\ln t + \frac{1}{\beta}\ln(\alpha\beta) \tag{12}$$

where α is the initial adsorption rate (mg g^{-1} min^{-1}), β is the adsorption constant (g mg^{-1}), and q_t is the adsorption capacity at any time t in mg g^{-1}. **Figure 15** shows a plot of linearization form of Elovich model. The slopes and intercepts of plots of q_t versus ln t were used to determine the constant β and the initial adsorption rate α.

3.4.4 Intra-particle diffusion model

When adsorbate transmits from solution into solid phase of absorbents, pore and intra-particle diffusion are often rate limiting in a batch reactor system. The intra-particle diffusion was explored by using the following equation suggested by Weber and Morris [33]:

$$q_t = K_{\text{int}} t^{1/2} + C \tag{13}$$

where the parameter q_t is the amount adsorbed at time t (mg g^{-1}), K_{int} is the intra-particle diffusion equation constant (mg g^{-1} min$^{-1/2}$), and t is the time. According to the Weber-Morris model, the plot of q_t, against $t^{1/2}$, should give a straight line, when diffusion plays a role in the sorption rate, and should cross the origin if intra-particle diffusion is the rate-determining step. The intra-particle diffusion parameters can be calculated from the slope and intercept of the linear plot given in **Figure 16**.

The kinetic parameters obtained by the sorption of heavy metal ions on PAZO nanocomposite are summarized in **Table 2**. It is found that the correlation coefficients R^2 for the pseudo-second-order kinetic model are 0.99 for 10, 50, and 100 mg L^{-1} Cr(VI) concentration at 50°C, respectively, is higher than the correlation coefficient obtained for other models. Also, the q_{cal} values obtained through this model 9.79, 25.46, and 38.44 mg g^{-1} are much closer to q_{exp} values 9.31, 25.11, and 37.89 mg g^{-1} for 10, 50, and 100 mg L^{-1} Cr(VI) solution. The pseudo-second-order kinetic model assumes that the rate limiting step may be chemical adsorption,

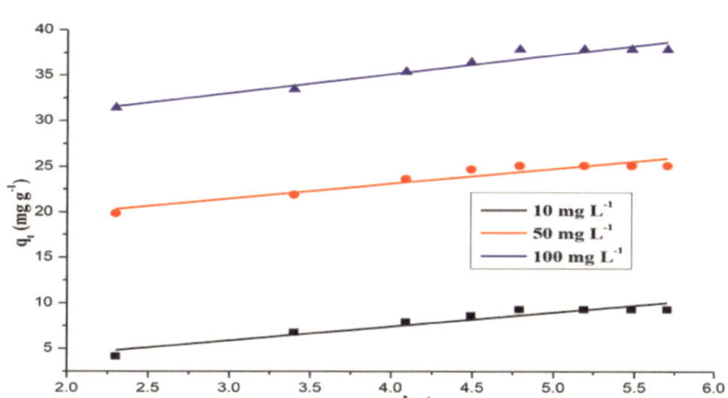

Figure 15.
Elovich model for Cr(VI) on PAZO at 10, 50, and 100 mg L^{-1} metal ion concentration (dose = 0.04 g and pH = 2).

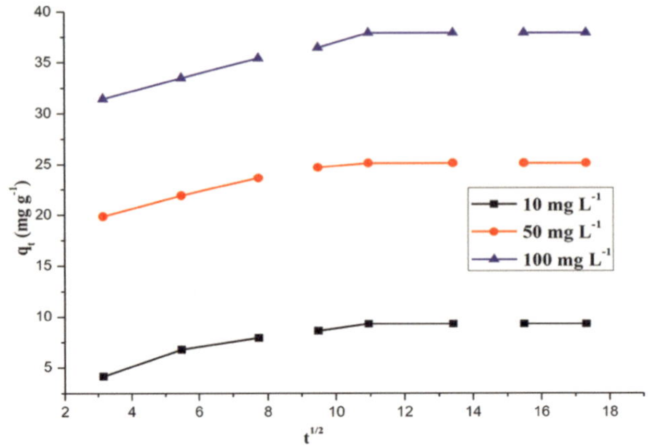

Figure 16.
Intra-particle diffusion model for Cr(VI) on PAZO at 10, 50, and 100 mg L^{-1} metal ion concentration (dose = 0.04 g and pH = 2).

Model	Parameters	Cr(VI)		
		30°C	40°C	50°C
Pseudo-first-order	qe (exp) (mg g^{-1})	9.31	25.11	37.89
	qe (cal) (mg g^{-1})	5.88	6.57	8.77
	k_1 (min^{-1})	0.02	0.02	0.02
	R^2	0.97	0.98	0.97
Pseudo-second-order	qe (exp) (mg g^{-1})	9.31	25.11	37.89
	qe (cal) (mg g^{-1})	9.79	25.46	38.44
	k_2 (g mg^{-1} min^{-1})	2.19×10^{-3}	2.17×10^{-3}	2.18×10^{-3}
	R^2	0.99	0.99	0.99
Elovich	α (mg g^{-1} min^{-1})	3.59	3.88	7.57
	β (g mg^{-1})	0.65	0.61	0.48
	R^2	0.90	0.89	0.93
Intra-particle	K_{int} (mg $g^{-1}$$min^{-1/2}$)	0.32	0.35	0.45
	C (mg g^{-1})	4.75	20.21	31.30
	R^2	0.69	0.70	0.79

Table 2.
Kinetic parameters for Cr(VI) removal by PAZO at 50°C.

and the adsorption behavior of PAZO may involve valence forces through sharing of electrons between transition metal cations and the PAZO. Because of the fact that diffusion and adsorption are often experimentally inseparable, the uptake of metal ions onto nanocomposite may be a complicate process including diffusion, coordinating bond formation or chemical reaction simultaneously. However, from the results obtained, it can be observed that good fits to the experimental data are obtained with pseudo-second-order model for the metal ion at all concentration ranges at 50°C.

3.5 Adsorption thermodynamics

To substantiate our prediction about the endothermic nature of the adsorption process, thermodynamic parameters such as Gibbs free energy change ($\Delta G°$), enthalpy change ($\Delta H°$), and entropy change ($\Delta S°$) were calculated using the Gibbs equation and the Van't Hoff equation, listed as follows:

$$\Delta G^o = -RT \ln K_c \tag{14}$$

$$\ln Kc = -\frac{\Delta H^o}{R} + \frac{\Delta S^o}{RT} \tag{15}$$

The gas constant R is defined by 8.3145 J mol^{-1} K^{-1}. K_c (C_{ad}/C_e) is the distribution coefficient; T is the temperature of the solution in Celsius. $\Delta G°$ and $\Delta S°$ were calculated from the slope and intercept of a plot of ln K_c as a function of $1/T$, as shown in **Figure 17**. The free energy change ($\Delta G°$) can be determined from the following equation:

$$\Delta G^o = \Delta H^o - T\Delta S^o \tag{16}$$

Thermodynamic parameters associated with the Cr(VI) adsorption by the nanocomposite are listed in **Table 3**. The positive value of $\Delta H°$ confirmed the endothermic nature of the adsorption process of Cr(VI) on PAZO. The values of $\Delta G°$ are all negative, and the negative value of $\Delta G°$ increases as the temperature increase from 30 to 50°C, which indicates that the Cr(VI) adsorption process of the nanocomposite is spontaneous and spontaneity increases with temperature [34]. The positive value of $\Delta S°$ revealed the increased randomness and an increase in the degrees of freedom at the adsorbent-solution interface during the immobilization of the heavy metal ions on the active sites of the adsorbent, which indicate the partial liberation of the salvation metal ions from solvent molecules before adsorption (liberation of water molecules from solvated-heavy metals), therefore, enabling commonness of randomness and spontaneity in the system.

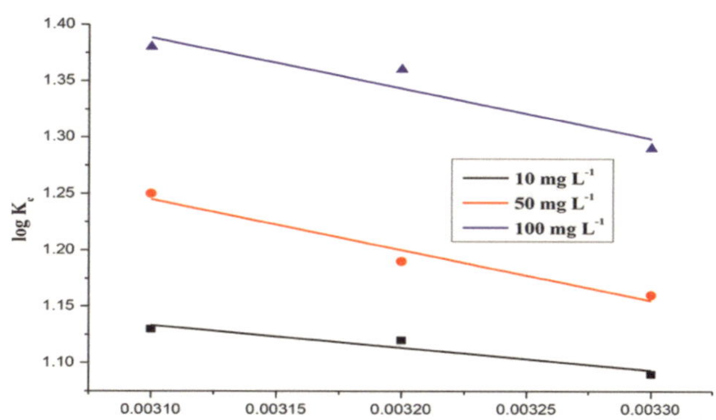

Figure 17.
Thermodynamic plot for removal of Cr(VI) on PAZO at 30, 40, and 50°C.

Concentration	ΔH° (KJ mol^{-1})	ΔS° (KJ mol^{-1} K^{-1})	ΔG° (KJ mol^{-1})		
			30°C	40°C	50°C
10 mg L^{-1}	1.66	0.014	−2.75	−2.89	−3.04
50 mg L^{-1}	3.74	0.021	−2.91	−3.13	−3.35
100 mg L^{-1}	3.74	0.023	−3.26	−3.50	−3.73

Table 3.
Thermodynamic parameters for Cr(VI) removal by PAZO.

4. Conclusion

In this chapter, polyaniline/zinc oxide nanocomposite was synthesized by oxidative free radical polymerization of aniline monomer in presence of zinc oxide nanoparticles. The material was characterized by various analytical techniques, such as FT-IR, XTD, TGA-DTG, SEM, and TEM. The nanocomposite material was further explored for the removal of Cr(VI) from aqueous solution. The effect of various adsorption parameters such as agitation time, solution pH, adsorbent dose, initial metal ion concentration, and temperature was observed and optimized by preliminary experiments. The adsorption of metal ions is highly pH-dependent and the maximum removal efficiencies and adsorption capacities of the selected metal ions were obtained at pH 2. The experimental data were tested using Langmuir, Freundlich, D-R and Temkin models and the data were best followed by Langmuir model. The maximum monolayer adsorption capacity was found to be 120.92 mg g^{-1} at 30°C, 134.22 mg g^{-1} at 40°C, and 139.47 mg g^{-1} at 50°C. All kinetic parameters suggest that the adsorption of metal ion by PAZO followed the second-order kinetics and chemisorption is the rate-determining step. The positive values of ΔH° and negative value of ΔG° indicate the adsorption process to be endothermic and spontaneous in nature.

Nomenclature

C_e	equilibrium adsorbate concentration in the solution (mg L^{-1})
C_O	initial adsorbate concentration in the solution (mg L^{-1})
C_{ab}	equilibrium concentration of adsorbate on the adsorbent surface (mg L^1)
q_e	amount of adsorbate adsorbed per unit mass of adsorbent at equilibrium or adsorption capacity (mg g^{-1})
m	amount of adsorbent (g)
DDW	double distilled water
AAS	atomic absorption spectrophotometer
FTIR	Fourier-transform infrared spectroscopy
SEM	scanning electron microscopy
EDX	energy dispersive X-ray analysis
TEM	transmission electron microscopy
XRD	X-ray diffraction
TGA	thermogravimetric analysis
q_e(exp)	adsorption capacity determined experimentally (mgg^{-1})
q_e(cal)	adsorption capacity calculated from model (mg g^{-1})
k_1	pseudo-first-order rate constant (min^{-1})
k_2	pseudo-second-order rate constant (g mg^{-1} min^{-1})
q_t	adsorption capacity at time t (mg g^{-1})
k_{id}	intra-particle diffusion rate constant (mg g^{-1} $min^{-1/2}$)
C	intra-particle constant related to thickness of boundary layer
A and B	Elovich constants (mg g^{-1})
K_c	equilibrium constant
b	Langmuir constant for energy of adsorption (L mg^{-1})
q_m	theoretical maximum monolayer adsorption capacity (mg g^{-1})
K_F	Freundlich isotherm constant (mg g^{-1})(L $mg^{-1})^{1/n}$
n	Freundlich exponent, dimensionless factor
K_T	Equilibrium binding constant (L mg^{-1})

B_T	Temkin constant related to heat of adsorption (J mol^{-1})
β	constant related to the adsorption energy (mol^2 kJ^{-2})
ε	Polanyi potential
E	mean free energy of adsorption (kJ mol^{-1})
ΔG	free energy change (kJ mol^{-1})
ΔH	enthalpy change (kJ mol^{-1})
ΔS	entropy change (kJ mol^{-1} K^{-1})
R	universal gas constant (J mol^{-1} K^{-1})
T	absolute temperature (K)
t	time (min)
V	volume of solution (liter)

Units

°C	degree celsius
K	kelvin
cm	centimeter
g	gram
mg	milligram
mg g^{-1}	milligram per gram
L	liter
ml	milliliter
mg L^{-1}	milligram per liter
µg L^{-1}	microgram per liter
M	molar
N	normality
h	hours
min	minute
J	joule
kJ	kilo joule
mol	mole
rpm	revolutions per minute
µg	microgram
Å	angstrom
nm	nanometer

Author details

Rais Ahmad
Environmental Research Laboratory, Department of Applied Chemistry, Aligarh Muslim University, Aligarh, India

*Address all correspondence to: rais45@rediffmail.com

References

[1] Li H, Li J, Chi Z, Ke W. Kinetic and equilibrium studies of Cr (III) removal from aqueous solution by IRN-77 cation-exchange resin. Procedia Environmental Sciences. 2012;**16**:646-655

[2] Chen S, Yue Q, Gao B, Xu X. Equilibrium and kinetic adsorption study of the adsorptive removal of Cr (VI) using modified wheat residue. Journal of Colloid and Interface Science. 2010;**349**:256-264

[3] Jain M, Garg VK, Kadirvelu K. Adsorption of hexavalent chromium from aqueous medium onto carbonaceous adsorbents prepared from waste biomass. Journal of Environmental Management. 2010;**91**:949-957

[4] Kocaoba S, Akcin G. Removal and recovery of chromium and chromium speciation with MINTEQA2. Talanta. 2002;**57**:23-30

[5] Pang Y, Zeng G, Tanga L, Zhang Y, Liu Y, Lei X, et al. Preparation and application of stability enhanced magnetic nanoparticles for rapid removal of Cr (VI). Chemical Engineering Journal. 2011;**175**:222-227

[6] AL-Othman ZA, Ali R, Naushad M. Hexavalent chromium removal from aqueous medium by activated carbon prepared from peanut shell: Adsorption kinetics, equilibrium and thermodynamic studies. Chemical Engineering Journal. 2012;**184**:238-247

[7] AL-Othman ZA, Ali R, Naushad M. Kinetic, equilibrium isotherm and thermodynamic studies of Cr (VI) adsorption onto low-cost adsorbent developed from peanut shell activated with phosphoric acid. Environmental Science and Pollution Research. 2013;**20**:3351-3365

[8] Shin K, Hong J, Jang J. Heavy metal ion adsorption behaviour in nitrogen-doped magnetic carbon nanoparticles: Isotherms and kinetic study. Journal of Hazardous Materials. 2011;**190**:36-44

[9] Noreen S, Bhatti HN, Nausheen S, Sadaf S, Ashfaq M. Batch and fixed bed adsorption study for the removal of Drimarine Black CL-B dye from aqueous solution using a lignocellulosic waste: A cost effective adsorbent. Industrial Crops and Products. 2013;**50**:568-579

[10] Noreen S, Bhatti HN, Nausheen S, Zahid M, Asim S. Biosorption of Drimarine Blue HF-RL using raw, pretreated and immobilized peanut hulls. Desalination and Water Treatment. 2014;**52**:7339-7353

[11] Tofighy MA, Mohammadi T. Adsorption of divalent heavy metal ions from water using carbon nanotube sheets. Journal of Hazardous Materials. 2011;**185**:140-147

[12] Gupta VK, Jain R, Saleh TA, Nayak A, Malathi S, Agarwal S. Equilibrium and thermodynamic studies on the removal and recovery of Safranine-T dye from industrial effluents. Separation Science and Technology. 2011;**46**:839-846

[13] Shahat A, Awual MR, Khaleque MA, Alam MZ, Naushad M, Chowdhury AMS. Large-pore diameter nano-adsorbent and its application for rapid lead (II) detection and removal from aqueous media. Chemical Engineering Journal. 2015;**273**:286-295

[14] Nausheen S, Bhatti HN, Furrukh Z, Sadaf S, Noreen S. Adsorptive removal of Drimarine Red HF-3D dye from aqueous solution using low-cost agricultural waste: Batch and column study. Chemistry and Ecology. 2015;**30**:376-392

[15] Ahmad R, Kumar R. Conducting polyaniline/iron oxide composite:

A novel adsorbent for the removal of amido black 10B. Journal of Chemical & Engineering Data. 2010;**55**:3489-3493

[16] Kanwal F, Rehman R, Anwar J, Saeed M. Batchwise removal of chromium (VI) by adsorption on novel synthesized polyaniline composites with various brans and isothermal modelling of equilibrium data. Journal of the Chemical Society of Pakistan. 2012;**34**: 1134-1139

[17] Zhu J, He H, Zhu L, Wen X, Deng F. Characterization of organic phases in them interlayer of montmorillonite using FTIR and 13C NMR. Journal of Colloid and Interface Science. 2005;**286**: 239-244

[18] Khan TA, Nazir M, Ali I, Kumar A. Removal of chromium(VI) from aqueous solution using guar gum-nano zinc oxide biocomposite adsorbent. Arabian Journal of Chemistry. 2013

[19] Vigneshwaran N, Kumar S, Kathe AA, Varadarajan PV, Prasad V. Functional finishing of cotton fabrics using zinc oxide-soluble starch nanocomposites. Nanotechnology. 2006;**17**:5087-5095

[20] Sedaghat S. Synthesis and characterization of new biocompatible copolymer: Chitosan-graft polyaniline. International Nano Letters. 2014;**4**:1-6

[21] Shi L, Wang X, Lu L, Yang X, Wu X. Preparation of TiO_2/polyaniline nanocomposite from a lyotropic liquid crystalline solution. Synthetic Metals. 2009;**159**:2525-2529

[22] Sharma AL, Saxena V, Annopoorni S, Malhotra BD. Synthesis and characterization of a copolymer: Poly (aniline-co-fluoroaniline). Journal of Applied Polymer Science. 2001;**81**: 1460-1466

[23] Yavuz AG, Dincturk-Atalay E, Uygun A, Gode F, Aslan E. A comparison study of adsorption of Cr (VI) from aqueous solutions onto alkyl-substituted polyaniline/chitosan composites. Desalination. 2011;**279**: 325-331

[24] Ajouyed O, Hurel C, Ammari M, Allal LB, Marmier N. Sorption of Cr (VI) onto natural iron and aluminum (oxy) hydroxides: Effects of pH, ionic strength and initial concentration. Journal of Hazardous Materials. 2010; **174**:616-622

[25] Monier M, Ayad DM, Sarhan AA. Adsorption of Cu (II), Hg (II), and Ni (II) ions by modified natural wool chelating fibers. Journal of Hazardous Materials. 2010;**176**:348-355

[26] Langmuir I. Adsorption of gaseous on plane surface of glass, mica and platinum. Journal of the American Chemical Society. 1918;**40**:1361-1403

[27] Freundlich HMF. Over the adsorption in solution. Zeitschrift für Physikalische Chemie. 1906;**57**:385-470

[28] Dubinin MM. The potential theory of adsorption of gases and vapours for adsorbents with energetically non-uniform surface. Chemical Reviews. 1960;**60**:235-266

[29] Temkin MJ, Pyzhev V. Kinetics of ammonia synthesis on promoted iron catalysts. Acta Physicochimica. 1940;**12**: 217-222

[30] Lagergren S. About the theory of so-called adsorption of soluble substances. The Hand. 1898;**24**:1-39

[31] Ho YS. Review of second-order models for adsorption systems. Journal of Hazardous Materials. 2006;**136**: 681-689

[32] El-Ashtoukhya E-SZ, Amina NK, Abdelwahab O. Removal of lead (II) and copper (II) from aqueous solution using

pomegranate peel as a new adsorbent. Desalination. 2008;**223**:162-173

[33] Weber WJ Jr, Morris JC. Kinetics of adsorption on carbon from solution. Journal of the Sanitary Engineering Division: American Society of Civil Engineers. 1963;**89**:31-59

[34] Ahmet Sari A, Mendil D, Tuzen M, Soylak M. Biosorption of palladium(II) from aqueous solution by moss (*Racomitrium lanuginosum*) biomass: Equilibrium, kinetic and thermodynamic studies. Journal of Hazardous Materials. 2009;**162**:874-879

Section 2

Mechanical and Wear Behaviour

Chapter 3

A Short Review on Al MMC with Reinforcement Addition Effect on Their Mechanical and Wear Behaviour

Vikas Verma and Alexandra Khvan

Abstract

Advanced mechanical and wear properties and applications of composites with bases of light weight metals have led to the need of aluminium (Al) metal matrix composites (MMCs). In today's time aluminium (Al) metal matrix composites (MMCs) are considered the most potential material for structural and functional applications. Composite materials with aluminium matrices are used in defence, aerospace, automotive and aviation, thermal management areas. Beneficial properties with reduced prices have enlarged their applications. To obtain desired physical and mechanical properties like high hardness, high strength, high stiffness, high wear, abrasion and corrosion resistance Al is reinforced with different metallic, non-metallic and ceramic elements. Al MMCs are used to make piston, connecting rod, engine cylinders, disc and drum brakes where wear has a great role in the functioning of these components as excessive wear of the mating components sometimes leads to catastrophic failures. Improvement of mechanical, especially tribological properties of hybrid composites were provided by the use of certain reinforce materials such as SiC, Al_2O_3 and graphite. Hence the present chapter presents a review on aluminium metal matrix composites (MMCs) reinforced with different particulate, whisker, fibres reinforcements highlighting their effect on physical, mechanical and wear behaviour of Al MMCs.

Keywords: aluminium (Al), metal matrix composite (MMC), stir casting, reinforcement, wear

1. Introduction

Aluminium (Al) amongst several metals is attractive due to its ductility, malleability, good conductivity, light weight, good strength and availability in abundance (8% of earth crust is aluminium). It combines with hard materials like ceramic and offer promising metal matrix composites (MMCs) with improved properties and hence finding wide range of industrial and structural applications including aerospace, automotive, marine and military [1–5]. For developing aluminium-based metal matrix composites various methods are applied by various researchers in liquid metallurgy routes for mass production. Reinforcement in aluminium metal matrix composites can be in particulate, whisker, continuous or discontinuous fibres.

Their addition to the base metal may vary in percentage resulting in improved properties. Composites having aluminium as base metal gives the following advantages: higher strength, improved stiffness, reduced density, survival at high temperature, high wear and corrosion resistance, improved damping capabilities [2].

For developing aluminium-based metal matrix composites various methods like powder metallurgy, spray decomposition, liquid metal infiltration, squeeze casting, mechanical alloying and compo casting are applied by various researchers in liquid metallurgy routes for mass production. Most common method use for processing of aluminium MMCs by powder metallurgy (PM). Via PM route aluminium MMCs can be prepared either by direct metal oxidation (DIMOX) or by reinforcements of particles in the matrix so as to achieve high density, high hardness and strength. In MMCs generally matrix component is more in quantity and reinforcement is a contrasting phase distributed in the matrix in order to reinforce it. The reinforcement rather than making a solid solution with the base matrix, it gets distributed all around it. When three constituents are present, it is called a hybrid composite. The aim of the reinforcement particles is to give high strength and stiffness to the composite and the aim of the matrix is to bind the reinforced particles together by virtue of its adhesive and cohesive nature and to transfer the load to and between reinforcements. In case of particle reinforced composites significant improvement is obtained in the mechanical properties in terms of strength, hardness and stiffness [6–8]. As a continuous phase, the matrix controls the interlaminar strength, elevated-temperature strength and transverse properties of the composite. The matrix holds reinforcing particles in the proper orientation and position so that they can carry the intended loads and distributes the loads evenly among the reinforcements so in a way matrix allows the strength of the reinforcements to be used to their full potential. The matrix also provides a vital inelastic response so that stress concentration are reduced and internal stresses are redistributed from broken reinforcements, reinforcements increase strength, decrease the coefficient of thermal expansion, and improve the wear resistance at a cost of a reduction in ductility and in fracture toughness [9]. Amongst the various methods employed to synthesize metal matrix composites, stir casting method is preferred and used for bulk production. The particular advantages of this process lie in its simplicity, cost effectiveness, flexibility and applicability to larger size components and mass production [10]. Selection of optimum parameters of stirring speed, stirring time, uniform feed rate of particles preheating temperature of the mould results in homogenous mixing and wetting of reinforced particles with base metal. It is seen that the cost of manufacturing of composite materials using a conventional casting method is about one third to half as that of competitive methods and, for high volume production, this cost is expected to reach the level of one-tenth [11]. In MMC's processing there are limitations with the conventional methods as conventionally produced composites are thermodynamically unstable when used at high temperature for longer time [12].

As it is known that today aluminium metal matrix composites are considered the most potential material for structural and functional applications and are finding versatile application in industries due to their price including defence, aerospace, automotive and thermal management areas, as well as in sports and recreation because of their unique isotropic properties of high strength, high stiffness, reduced density (weight), high wear, abrasion and corrosion resistance and improved high temperature properties. These properties are limited in conventional alloys [1–5]. Some of the applications of Al MMCs are shown in **Figure 1** [13]. It is reported that in aluminium-based metal matrix composites fabrication aluminium is reinforced with different reinforcing material like MgO, SiC, MnO, Al_2O_3 which give high mechanical properties to these composites like hardness, fracture toughness and reduced density (weight). Al MMCs consist of hard particles like SiC, WC,

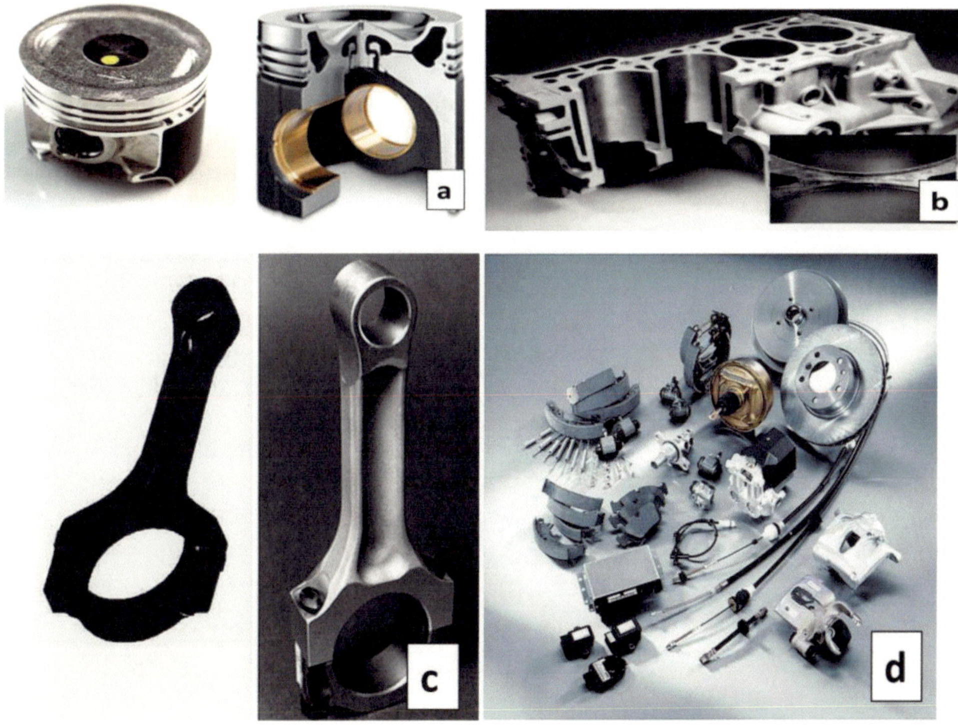

Figure 1.
(a) Piston, (b) engine with cylinder barrel, (c) piston connecting rod, (d) brake system made of aluminium (Al) metal matrix composites (MMCs) [13].

Al_2O_3, etc. and these particles make the aluminium matrix plastically constrained which improves its high temperature properties and they give superior mechanical and wear resistant properties [14].

2. Physical and mechanical properties of aluminium-based metal matrix composites

Researchers have reinforced Al matrix with different metallic, non-metallic and ceramic elements to have desired physical and mechanical properties. Stir casting given by Ray [15] is the best liquid state fabrication technique through which metal matrix composites can be successfully processed. In this method reinforcements are dispersed in molten metal matrix by mechanical stirring as shown in **Figure 2**. Al-Al_2O_3 (MnO_2) hybrid MMCs were processed *via* stir casting technique by adding varying wt.% of Al_2O_3 and MnO_2 particles to the Al melt at high temperature and stirred for uniform distribution of the particles in the melt. At high temperature MnO_2 particles react with molten aluminium and get reduced to metallic manganese, and retain in the matrix of molten aluminium [15, 16] Al-Al_2O_3 composite were developed by direct metal oxidation (DIMOX) and hybrid composite by using both ex situ and *in situ* approaches together by a dispersing powder mixture of Al_2O_3 and MnO_2 in a ratio of 1:1, through stir casting route in aluminium matrix [17]. Al-Al_2O_3 composites were synthesized successfully under ambient conditions and it was found that the particles increased as the holding time and the number of stirrings was increased resulting in improved mechanical properties. Hence, oxidation, which is known as a major problem in composite preparation by liquid metallurgy route, can be used for preparation of

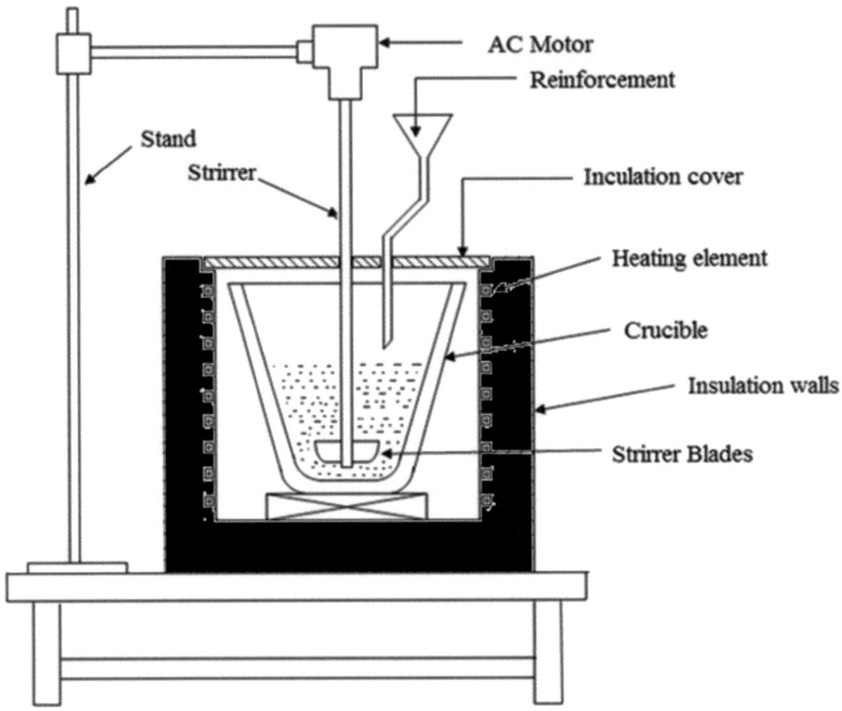

Figure 2.
Stir casting set up schematic diagram.

composites and for improving the properties. In hybrid composites the mechanical properties as well as particle distribution improved as Al_2O_3 + MnO_2 were introduced and also with their increasing wt.%. It is reported that reinforcing with 60 vol% Al reinforcement elastic modulus of pure aluminium increase from 70 to 240 GPa and leads to decrease to the coefficient of expansion from 24 to 7 ppm/°C [18]. Jokhio et al. [19] found that the addition of 2.77 wt.% Mg contents in Al matrix increases wettability, reduces porosity and develops high bonding with Al_2O_3. Wahab et al. [20] in their work found an increase in hardness value (from 44 to 94 HV at 100 g load) of Al-Si alloy when reinforced with 0–10 increasing wt.% of AlN. Tzamtzis et al. [21] found that the addition of 1 wt.% of Mg in A356/SiCp composites increases wettability and its further increases results in agglomeration of particles which increases viscosity resulting in non-uniform distribution of particles in the matrix. Saheb [22] found that the addition of varying weight fraction of silicon carbide, graphite and alumina in Al matrix, the hardness value of Al MMCs increases, maximum hardness of 45.5 and 74 BHN have been obtained with addition of 25% weight fraction of SiC and at 4% weight fraction of graphite respectively. It is observed in mechanical and microstructural behaviour of Al-7075/B_4C composites with constant weight of B_4C processed at varying temperatures (450–540°C) that due to the formation of MgO at higher sintering temperature above 530°C led to decrease in hardness and bending strength [23]. Increase of 81 and 37% in yield strength and ultimate tensile strength is observed in nano-Al_2O_3/2024 composites fabricated by solid-liquid mixed casting combined with ultrasonic treatment [24]. The effect of graphite particle on Al6082 metal matrix composite is investigated by Sharma et al. [25]. Agglomerations of graphite particles at some points were found with presence of large impurities due to non-uniform distribution. Hardness is reduced to 11.1% with 12 wt.% Graphite due to brittle nature of Gr reinforcement particles. Summarising, addition of different reinforcement have resulted in different microstructural features leading to distinct physical and mechanical properties in aluminium-based metal matrix composites (Al MMCs).

3. Wear behaviour of aluminium-based metal matrix composites

Now a days, aluminium-based metal matrix composites (Al MMCs) are used in making of piston, connecting rod, contactors, where sliding is an important factor [26]. Excessive wear of the mating components sometimes leads to catastrophic failures [27]. So study of wear properties of Al MMCs has become the need of time. Wear tests are generally conducted on ball/pin wear tester, schematic diagram as represented in **Figure 3**. Wear properties of many MMCs having continuous and discontinuous reinforcements like Al_2O_3, MnO_2, SiC, graphite, mica, glass, graphite and others have been reported [28–30].

There has been increasing interest in composites and many researchers contributing their work in the area of wear analysis of aluminium composites, cermets, ceramics [30–40]. Umanath et al. [30], examined the effect of SiC and Al_2O_3 on dry sliding wear behaviour of Al6061 hybrid composites prepared by stir casting method, results showed that with increase in the volume content, wear decreases due to the presence of hard oxide particles. Suresh et al. [39] investigated the wear behaviour of Al6061 reinforced with Al_2O_3 and graphite by keeping 2 wt.% graphite constant and Al_2O_3 content is varied 2–8 wt.%. The reinforcement of Al_2O_3 and graphite improved the tribological behaviour and caused reduction in the wear of Al6061 composites. The wear decreased with the increase of speed and aluminium oxide percentage. Basavarajappa and Chandramohan [40], worked on the dry sliding wear behaviour of Al2219 reinforced with SiC (0–15 wt.%). Results shows that 15% SiC reinforced composites have better wear resistance then other composites. Raghavendra and Ramamurthy [41], examined the influence of particle size and volume fraction on wear behaviour of Al7075 alloy reinforced with Al_2O_3 particles, size is varied 100–200 microns and the volume fraction is varied 3–12 wt.%. Results showed that hardness

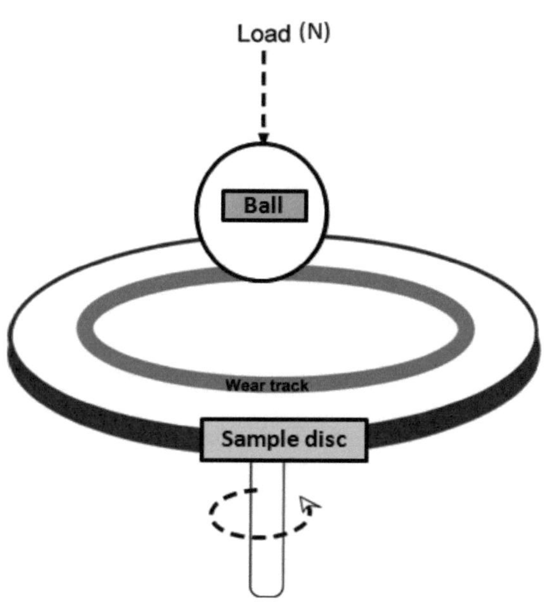

Figure 3.
Schematic diagram for sliding wear ball on disc tests.

increased with decrease in particle size and the wear rate was reduced with reduction in particle size. Increase in volume fraction reduced the wear and coefficient of friction. Vivekanandan et al. [42], investigated the wear resistance by varying load on the fly ash reinforced composites. Fly ash was added to the aluminium alloy and fabricated by stir casting method. At varying load it was noticed that wear rate of composites was less than the pure alloy at all loads. Kumar et al. [43], worked on the mechanical and wear behaviour of aluminium-fly ash composites formed by stir casting method. It was found that the hardness of composites increased with increase in addition of fly ash. Addition of fly ash shows improvement in the strength of composites. Strengthening of composites is due to dispersion and reinforcement. Both the wear rate and frictional force decreased with the adding of fly ash in Al6063 alloy. The aforementioned literatures show that various researchers have attempted to improve the properties of Al alloy by adding different alloying elements.

Al_2O_3 alloy when reinforced with 20 wt.% of alumina gives better wear resistance properties [44]. Al6061-alumina fibre composites abrasive wear rate is reported to be very less than the matrix alloy and is reported to have better wear resistance almost six times the matrix alloy [44]. The reason attributes to it is due to the addition of hard ceramic particles. Wear rate of Al7091 alloy and Al7091-SiC composites have almost same wear rate at 1.2 m/s sliding velocity whereas at increasing sliding velocity composites show less wear than un reinforced matrix [45–47]. TiO_2 as reinforcement in Al alloys give high mechanical properties as hardness and superior corrosion resistance [11]. Al6061 is considered as candidate material to prepare MMCs owing to its better formability characteristics and option of modification of the strength of composites by adopting optimal heat treatment [26]. Dinaharan et al. [48] fabricated aluminium alloy Al6061 reinforced with ZrB_2 particles (10 wt.%) and found ZrB_2 particles into the aluminium matrix improved tensile strength and wear resistance but reduced ductility and corrosion resistance. The wear resistance was measured using a pin-on-disc wear apparatus at room temperature according to ASTM G99-04 standard under dry sliding conditions. The polished surface of the pin of 6 × 6 × 50 mm was slide on a hardened chromium steel disc. The test was carried out at a sliding velocity of 15 m/s, normal force of 25 N and sliding distance of 2500 m. Wear resistance for Al6061 and Al6061/10 ZrB_2 is found to be 182.48 and 377.51 m/mm^3. The pitting corrosion rate was measured using potentiodynamic anodic polarisation technique as per ASTM G5 (ACM Gill-5500) at room temperature and found that 0.0230 and 0.1746 mm/year corrosion rate for Al6061 and Al6061/10 ZrB_2. Lus et al. [49] investigated the wear properties of *in situ* A380-Mg_2Si alloy squeeze cast composites. Wear tests were carried out on pin-on-drum type equipment where a rotating disk (34 rev min^{-1}) with 120 grid sand paper. Samples were placed vertically with loads such as 20, 25, 30, 35 and 40 N. The travel distances were selected to be 20, 40, 60 and 80 m [49]. Wear tests were basically carried out in two folds. The conditions of the first set of experiments were constant load of 20 N with changing travel distances of 20, 40, 60 and 80 m. As seen in the volumetric loss was linear where the samples cast in die without any pressure had the lowest loss of 0.0565 cm^3 and the casting with the highest pressure (i.e., 40 MPa) had the highest weight loss of 0.127 cm^3 at 80 m sliding distance. In the second set of experiments, the travel distance was kept constant at 20 m and the load was increased from 20 to 40 N by 5 N increments. This time, the volumetric loss was exponential the parts cast under 40 MPa had the highest volumetric loss of 0.26 cm^3 [49]. Volumetric loss and wear rate of *in situ* Mg_2Si-A380 composite increases exponentially when the load is increased. The hardness measurements were carried out on the matrix of the samples by using Micro hardness tester (HV10) and the hardness values were 138, 108, 87 HV10 for 0, 30, and 40 MPa squeeze pressures, respectively [49].

The effect of external ultrasonic treatment during solidification of a casted hypereutectic Al-Si (18% Si) alloy is studied by Unal et al. [50] and found that it has favourably affected the hardness and provided an increase of 15–20%. From the pin on disc wear tests performed under 67 N and with 1250 m sliding distance, it was revealed that the ultrasonic treated and non-treated samples exhibited similar amounts of weight loss [50]. Yamanoglu et al. [51] studied the effect of nickel (1–5 wt.%) on microstructure and pin on disc wear behaviour of pure aluminium against steel and alumina counter faces. The dry sliding wear response of the Al-*x*Ni alloys against steel and alumina counter faces was investigated [51]. The results showed that the hardness of the alloys increases with increasing nickel content; it is 46.2 HV3 at 5% Ni.

Severe wear damage was observed at low and high nickel contents. Maximum wear resistance was obtained with the addition of 3 wt.% nickel to the pure aluminium under both loads and against both counter faces. The wear resistance of the alloys increased with increasing nickel content up to 3 wt.% Ni and tended to decrease >3 wt.% Ni. The wear rate of the Al-*x*Ni alloys increased with increasing applied normal load [51]. One of the disadvantages with Al is that it leads to corrosion. It is preferentially in use in marine applications and to protect it from corrosion anti corrosive coatings are done on it like that of zinc chromate. The harmful effect of zinc chromate is that it is toxic in nature and is hazardous to the environment. Its replacement is done with cerium oxide which is available in nature abundantly and is less toxic. Wear study of Al_2O_3 and B_4C reinforced with Al5083 alloy matrix processed by stir casting method shows that shows that at varying load increasing wt.% of B_4C from 8 to 12% has increased the wear rate also [52]. Iacob et al. [53] observed the wear rate of $Al/Al_2O_3/Gr$ hybrid composites and found low wear rate as Al_2O_3 and Gr acted as load bearing elements and solid lubricant. Also, there is increase in micro hardness due to Al_2O_3 particles. Ganesh et al. [54] studied the mechanical and wear behaviour of Al2219 alloy—SiC composites and found that higher wt.% of SiC (20%) and sintering temperature (600°C) resulted in high hardness and reduced wear at varying wear load and speed. Using Taguchi Technique Al-Al_2O_3 composites wear behaviour is studied and optimised by Baradeswaran et al. [55]. Taguchi analytical and graphical results shows minimum wear loss at 6 wt.% of Al_2O_3 at 10 N load and sliding distance of 400 m as optimum combination. Wear tests were conducted for 10, 20, 30, and 40 N load and sliding distance of 1200 m with regular interval of 200 m at 0.6 m/s sliding speed.

Summarising, literature survey shows that Al base MMCs are of huge need as they are used for various applications as for making different machine components as heavy duty pistons, aircraft generator housings, air cooled cylinder heads, engine crankcases, petrol and oil tanks, oil pans, water cooled cylinder heads, rear axle housings, flywheel housings, automotive transmission cases, oil pans, rear axle housings, brackets, water cooled cylinder blocks, various fittings and pump bodies, air brake castings, gear cases, air cooled cylinder heads, air brake castings, gear cases, air cooled cylinder heads, Internal combustion engine pistons and blocks, cylinder bodies for compressors, pumps and brakes which gets degraded with passage of time due to either wear (abrasive, adhesive) or corrosion so its becomes very essential to study their wear properties [56].

4. Conclusions

Present industrial developments are associated with materials having advantageous physical, mechanical and wear characteristics that can achieve technological needs. Aluminium and its composites are best suited materials as have better properties than unreinforced materials. Beneficial properties with reduced prices have enlarged their applications. Al MMCs are used in defence, aerospace, automotive, aviation, thermal management areas in engine pistons, cylinders barrel, connection rods, elements of vehicles braking systems because of their unique properties of high hardness, high strength, high stiffness, high wear, abrasion and corrosion resistance.

Author details

Vikas Verma* and Alexandra Khvan
Thermochemistry of Materials Scientific Research Centre, National University of Science and Technology "MISIS", Moscow, Russia

*Address all correspondence to: vikasverma.iitr@rediffmail.com

References

[1] Tiryakioglu M, Campbell J. Guidelines for designing metal casting research: application to aluminium alloy castings. International Journal of Cast Metals Research. 2007;**20**:25-29

[2] Clyne TW, Withers PJ. An introduction to Metal Matrix Composites. Cambridge University Press; 1993. p. 293

[3] Sajjadi SA, Ezatpour HR, Beygi H. Microstructure and mechanical properties of Al–Al_2O_3 micro and nano composites fabricated by stir casting. Materials Science and Engineering A. 2011;**528**:8765-8771

[4] Miller WS, Zhuang L, Bottema J, Wittebrood AJ, Smet PD, Haszler A, et al. Recent development in aluminium alloys for the automotive industry. Materials Science and Engineering A. 2000;**280**:37-49

[5] Brown KR, Venie MS, Woods RA. The increasing use of aluminium in automotive applications. Journal of the Minerals, Metals and Materials Society. 1995;**47**:20-23

[6] Kok M. Production and mechanical properties of Al_2O_3 particle-reinforced 2024 aluminium alloy composites. Journal of Materials Processing Technology. 2005;**161**:381-387

[7] Rosso M. Ceramic and metal matrix composites: Routes and properties. Journal of Materials Processing Technology. 2006;**175**:364-375

[8] Xiu ZY, Chen GQ, Wang XF, Wu GH, Liu YM, Yang WS. Microstructure and performance of Al-Si alloy with high Si content by high temperature diffusion treatment. Transactions of Nonferrous Metals Society of China. 2010;**20**:2134-2138

[9] Kuhn H, Medlin D. Mechanical Testing and Evaluation, American Society of Metals ASM Handbook ASM International. Handbook Committee: USA. 2000;**8**

[10] Kumar B, Menghani JV. Aluminium-based metal matrix composites by stir casting: A literature review. International Journal of Materials Engineering Innovation. 2016;**7**:1-14

[11] Surappa MK, Rohatgi PK. Preparation and properties of cast aluminium-cerarnic particle composites. Journal of Materials Science. 1981;**16**:983-993

[12] Xiaoming W, Animesh J, Rik B. Fabrication of Al_3Ti particle reinforced aluminium alloy metal-matrix composites. Materials Science and Engineering A. 2004;**364**:339-345

[13] Stojanovic B, Ivanovic L. Application of aluminium hybrid composites in automotive industry. Technical Journal. 2015;**22**:247-251

[14] Rohatgi PK, Yarandi FM, Liu Y In: Fishman SG, Dhingra AK (Eds.). Proceedings of International Symposium on Advances in Cast Reinforced Metal Composites, ASM International Publication, Materials Park, OH. 1988;249

[15] Ray S. Synthesis of cast metal matrix particulate composites. Journal of Materials Science. 1993;**28**:5397-5413

[16] Abdulhaqq AH, Ghosh PK, Jain SC, Ray S. Processing, microstructure, and mechanical properties of cast in-situ Al (Mg, Mn)-Al_2O_3 (MnO_2) composite. Metallurgical and Materials Transactions A. 2005;**36A**:2211-2223

[17] Verma V, Kumar P, Mittal K, Chauhan S, Tewari PC. Microstructure and mechanical behavior characterization of Al-Al_2O_3 MMC processed by DIMOX and Al-Al_2O_3/MnO_2 MMC processed via stir casting

route. International Journal of Materials Engineering Innovation. 2016;7:219-235

[18] Surappa MK. Aluminium matrix composites: Challenges and opportunities. Sadhana. 2003;**28**:319-334

[19] Jokhio MH, Panhwar MI, Unar MA. Manufacturing of Aluminum Composite Material Using Stir Casting Process, Mehran University Research Journal of Engineering & Technology.2011;**30**:53-64

[20] Wahab MN, Daud AR, Ghazali MJ. Preparation and characterization of stir cast-aluminum nitride reinforced aluminum metal matrix composites. International Journal of Mechanical and Materials Engineering. 2009;**4**:115-117

[21] Tzamtzis S, Barekar NS, Babu NH, Patel J, Dhindaw BK, Fan Z. Processing of advanced Al/SiC particulate metal matrix composites under intensive shearing: A novel Rheo-process. Composites Part A: Applied Science and Manufacturing. 2009;**40**:144-151

[22] Saheb DA. Aluminum silicon carbide and aluminum graphite particulate composites. ARPN Journal of Engineering and Applied Sciences. 2011;**6**:41-46

[23] Wu C, Fang P, Luo G, Chen F, Shen Q, Zhan L, et al. Effect of plasma activated sintering parameters on microstructure and mechanical properties of Al-7075/B_4C composites. Journal of Alloys and Compounds. 2014;**615**:276-282

[24] Su H, Gao W, Feng Z, Lu Z. Processing, microstructure and tensile properties of nano-sized Al_2O_3 particle reinforced aluminum matrix composites. Materials and Design. 2012;**36**:590-596

[25] Sharma P, Khanduja D, Sharma S. Dry sliding wear investigation of Al6082/Gr metal matrix composites by response surface methodology. Journal of Materials Research and Technology. 2016;**5**:29-36

[26] Sathyanarayana KG, Pillai RM, Pai BC, Kestursatya M, Rohatgi PK, Kim JK. Developments in cast metal matrix composites over last three and half decades, In: Dwarakadas ES and Krishnadas Nair GS, eds., Proceedings of the Third International Conference on Advances in Composites. Bangalore; 2002. pp. 753-763

[27] Ramesh CS, Seshadri SK, Iyer KJLA. Survey on aspects of wear of metals. Indian Journal of Technology. 1991;**29**:179-185

[28] Ramesh CS, Noor Ahmed R, Safiualla M. Strength and wear properties of cast copper-TiO_2-boric acid hybrid composites, in: V. C. Venkatesh and S. Mirdha, eds., Proceedings of the International Conference ICMAT, Kaula Lumpur, 2004:836-839

[29] Shashishankar A, Krishna M, Chandrasekhara Murthy CS. A study on sliding behaviour of flyash reinforced aluminium 7075 alloy composites, in: E. S. Dwarakadas and C. G. Krishnadas Nair, eds., Proceedings of the Third International Conference on Advances in Composites. Bangalore; 2002. pp. 583-589

[30] Umanath K, Selvamani ST, Kumar K, Sabarikreeshwaran R. Dry sliding wear behavior of AA6061-T6 reinforced SiC and Al_2O_3 particulate hybrid composites. Procedia Engineering. 2014;**97**:694-702

[31] Verma V, Manoj Kumar BV. Effects of binders (Ni-Co) and ternary carbide (TaC) on friction and wear behaviour of Ti (CN) based cermets, Advances in High Temperature Ceramic Matrix Composites and Materials for Sustainable Development, Ceramic Transactions, Wiley, CCLXIII, 2017. pp. 353-364

[32] Verma V, Manoj Kumar BV. Sliding wear behavior of SPS processed TaC containing Ti(CN)-WC-Ni/Co cermets against silicon carbide. Wear. 2017;**376-377**:1570-1579

[33] Verma V, Manoj Kumar BV. Tribological characteristics of conventionally sintered TiCN-WC-Ni/Co cermets against cemented carbide. Ceramics International. 2017;**43**:368-375

[34] Verma, V, Manoj Kumar BV. TaC-containing Ti(CN)-WC-Ni/Co cermets for the improved machining performance, In: E. S. Dwarakadas and C. G. Krishnadas Nair, eds., Sips 2017—Sustainable Industrial Processing Summit and Exhibition, Cancun, Mexico. 2017;**8**:210-222

[35] Verma V, Manoj KBV. Processing of alumina based composites via conventional sintering and their characterization. Materials and Manufacturing Processes. 2017;**32**:21-26

[36] Verma V, Manoj Kumar BV, Kang S. Sliding wear behavior of TaC-containing Ti(CN)-WC-Ni/Co cermets. International Journal of Applied Ceramic Technology. 2016;**13**:1033-1042

[37] Verma V, Manoj KBV. Tribological behavior of TiCN based cermets against steel and cemented carbide. Materials Today: Proceedings. 2016;**3**:3130-3136

[38] Verma V, Manoj KBV. Synthesis, microstructure and mechanical properties of $Al_2O_3/ZrO_2/CeO_2$ composites with addition of nickel and titania processed by conventional sintering. Materials Today: Proceedings. 2017;**4**:3062-3071

[39] Suresh R, Kumar M. Investigation of tribological behavior and its relation with processing and microstructures of Al6061 metal matrix composites. International Journal of Research in Engineering & Technology. 2013;**1**:91-104

[40] Basavarajappa S, Chandramohan G. Dry sliding wear behaviour of hybrid metal matrix composites. Materials Science. 2005;**11**:253-257

[41] Raghavendra N, Ramamurthy VS. Effect of particle size and weight fraction of alumina reinforcement on wear behaviour of aluminium metal matrix composites. International Journal of Innovative Research in Science Engineering and Technology. 2014;**3**:1191-1198

[42] Vivekanandan P, Anand S, Thanikaselam A. Investigation on wear resistance of varying load of particulate lignite fly ash reinforced aluminum alloy 6063 composites. International Journal of Emerging Technology and Advanced Engineering. 2012;**2**:145-148

[43] Kumar MR, Priyana MS, Mani A. Investigation of mechanical and wear properties of aluminium-fly ash composite material produced by stir casting method. 2014;**5**:1261-1269

[44] Hosking FM, Folgar Poritillo F, Wunderlein R, Mehrabion R. Composite of Al alloys fabrication and wear behaviour. Journal of Materials Science. 1982;**17**(2):477-498. DOI: 10.107/bf00591483

[45] Wang A, Hutchings IM. Wear of alumina fiber: Aluminum metal matrix composite by two body abrasion. Materials Science and Technology. 1989;**5**(1):71-76. DOI: 10.1179/026708389790337503

[46] Wang A, Rack HJ. Transition wear behavior of SiC-particulate and SiC-whisker-reinforced 7091 al metal matrix composite. Materials Science and Engineering A. 1991;**147**(2):211-224. DOI: 10.1016/0921-5093(91)90848-h

[47] Ramesh CS, Anwar Khan AR, Ramachandra A. Heat treatment of al6061-10 wt% sic composites. In: Anwarul haque AFM, Ahmed M,

Mustafizul Karim AN, Dhar NR, Begum S, editors. Proceedings of the International Conference on Manufacturing, ICM Dhaka. 2002. pp. 21-28

[48] Dinaharan I, Murugan N. Microstructure and some properties of aluminium alloy Al6061 reinforced in situ formed zirconium diboride particulate stir cast composite. International Journal of Cast Metals Research. 2014;**27**(2):115-121

[49] Murat Lus H, Ozer G, Altug Guler K, Erzi E, Dispinar D. Wear properties of squeeze cast in situ Mg_2Si–A380 alloy. International Journal of Cast Metals Research. 2015;**28**(1):59-64

[50] Unal N, Camurlu HE, Koçak S, Düztepe G. Effect of external ultrasonic treatment on hypereutectic cast aluminium–silicon alloy. International Journal of Cast Metals Research. 2012;**25**(4):246-250

[51] Yamanoglu R, Karakulak E, Zeren M, Koç FG. Effect of nickel on microstructure and wear behaviour of pure aluminium against steel and alumina counterfaces. International Journal of Cast Metals Research. 2013;**26**(5):289-295

[52] Hariprasad T, Varatharajan K, Ravi S. Wear characteristics of B_4C and Al_2O_3 reinforced with Al 5083 metal matrix based hybrid composite. Procedia Engineering. 2014;**97**:925-929

[53] Iacob G, Ghica VG, Buzatu M. Studies on wear rate and micro-hardness of the Al/Al_2O_3/Gr hybrid composites produced via powder metallurgy. Composites Part B: Engineering. 2014

[54] Ganesh R, Subbiah R, Chandrasekaran K. Dry sliding wear behavior of powder metallurgy aluminium matrix composite. Materials Today: Proceedings. 2015;**2**:1441-1449

[55] Baradeswaran A, Elayaperumal A, Franklin Issac R. A statistical analysis of optimization of wear behaviour of Al-Al_2O_3 composites using Taguchi technique. Procedia Engineering. 2013;**64**:973-982

[56] Culliton D, Betts AJ, Kennedy D. Impact of intermetallic precipitates on the tribological and/or corrosion performance of cast aluminium alloys: A short review. International Journal of Cast Metals Research. 2013;**26**(2):65-71

Section 3

Material's Models Developing

Chapter 4

An Alternative Framework for Developing Material Models for Finite-Strain Elastoplasticity

Ladislav Écsi, Pavel Élesztős, Róbert Jerábek, Roland Jančo and Branislav Hučko

Abstract

Contemporary plasticity theories and their related material models for finite deformations are either based on additive decomposition of a strain-rate tensor or on multiplicative decomposition of a deformation gradient tensor into an elastic part and a plastic part. From the standpoint of the nonlinear continuum mechanics, the former theories, which are used to model hypoelastic-plastic materials, are rather incomplete theories, while the latter theories, which are used to model hyperelastic-plastic materials, are not even continuum-based theories, while none of their related material models are thermodynamically consistent. Recently, a nonlinear continuum theory for finite deformations of elastoplastic media was proposed, which allows for the development of objective and thermodynamically consistent material models. Therefore, the analysis results of the models are independent of the description and the particularities of their mathematical formulation. Here by the description we mean total or updated Lagrangian description and by the particularities of formulation, the ability to describe the model in various stress spaces using internal mechanical power conjugate stress measures and strain rates. In this chapter, an alternative framework for developing objective and thermodynamically consistent hypoelastic-plastic- and hyperelastic-plastic-based material models is presented using the first nonlinear continuum theory of finite deformations of elastoplastic media.

Keywords: nonlinear continuum theory for finite deformations of elastoplastic media, objective and thermodynamically consistent formulation, J_2 generalised plasticity with isotropic hardening, hypoelastic-plastic- and hyperelastic-plastic-based material models with internal damping

1. Introduction

There are two types of phenomenological flow plasticity theories and their related material models used at present to model plastic behaviour of deformable bodies within the framework of finite-strain elastoplasticity. The first type of theories are considered to be ad hoc extensions of small-strain flow plasticity theories into the area of finite deformations to describe materials, in which small elastic deformations are accompanied by finite inelastic deformations during the deformation process. They are based on additive decomposition of a strain-rate tensor into an elastic part and a plastic part to describe the plastic flow in the material. As an example of such material is ductile metal, which at present is modelled mainly by a kind of a hypoelastic-plastic-based material model, whose constitutive equation does not have a form in terms of a finite-strain measure. Without a need for completeness, let us just mention a few comprehensive studies in technical literature, such as [1–4], where detailed descriptions of the most frequently used contemporary hypoelastic-plastic-based material models are presented.

The second type of flow plasticity theories, in which multiplicative decomposition of a deformation gradient tensor into an elastic part and a plastic part is used to describe the plastic flow in the material, is based on the theory of single-crystal plasticity [5–7]. The theories and their related material models, which are now considered as 'proper material models' to model plastic behaviour of the deformable body, assume that the intermediate configuration of the body is stress-free [1] or at least locally unstressed [4]. As a result, there cannot exist a deformation or strain tensor field that meets the conditions of compatibility [4]. Therefore these theories treat the kinematics of motion differently between the initial and current or intermediate configurations of the body. This means that the motion, the displacement field and the deformation gradient, all of which have an exact physical meaning in continuum mechanics, are considered in accordance with the continuum theory between the initial and current configurations of the body, but not between the configurations where one is an intermediate configuration. Here the motion and the displacement fields are disregarded, and as a result, the deformation gradient loses its physical meaning. Moreover, it should also be noted that the assumption of an unstressed intermediate configuration is not compatible with the theory of nonlinear continuum mechanics, as it violates proper stress transformations, resulting from the invariance of the internal mechanical power, when switching from one stress space to the other in any configuration of the body. As a result, contemporary multiplicative plasticity theories and their related material models in realty are not continuum-based.

Hypoelastic-plastic- and hyperelastic-plastic-based material models have been the subject of study over recent decades, and there are a few issues to be concerned about when the models are used in numerical analyses. These include energy accumulation and residual stresses along a closed elastic strain path in the case of hypoelastic-plastic-based material models in which the Jaumann rate is used to calculate the Cauchy stress tensor [8], residual stress accumulation up to unacceptable values during multiple loading cycles along a closed elastic strain path using the Jaumann rate and a few other rates [9] or shear oscillation in finite shearing problems [10, 11]. The aforementioned problems however can be eliminated by replacing the Jaumann rate with the Green-Naghdi or Truesdell rate in the formulation of the models [4]. Simo and Pistner showed that employing a constant spatial elasticity tensor in objective stress integration is not compatible with elasticity and that such models in fact fail to define the elastic material [12]. Equivalent rate descriptions of hyperelastic-based models in terms of different strain measures have been thoroughly discussed by Perić, who also showed that the Jaumann rate- and the Green-Naghdi rate-based models provide different levels of approximation to problems governed by the logarithmic strain-based Hencky hyperelastic law [4, 13]. We will show herein that all of the above problems actually resulted from the fact that the related material models use thermodynamically inconsistent formulation.

The aim of this chapter is to present an alternative framework for developing objective and thermodynamically consistent hypoelastic-plastic- and hyperelastic-plastic-based material models using the first nonlinear continuum theory for finite deformations of elastoplastic media [14]. We will show that the strain-rate tensor additive decomposition-based theories are in reality finite-strain theories, but they are constrained when the plastic flow in them is defined in terms of a Cauchy's stress tensor-based yield surface in the current configuration of the body, while contemporary deformation gradient multiplicative split-based theories are not even continuum-based theories. Moreover, none of their related material models is thermodynamically consistent. In addition to this, we will show that all flow plasticity theories are just variants of the nonlinear continuum theory for finite deformations of elastoplastic media presented in this chapter, using the additive decomposition of a Lagrangian displacement field into an elastic part and a plastic part. Eventually, we will demonstrate the theory in numerical experiments using simple hypoelastic-plastic- and hyperelastic-plastic-based material models with internal damping and then briefly discuss their analysis results.

2. Theory

The Lagrangian description is used to describe the kinematics of motion and constitutive and evolution equations of the material of a deformable body. Though a single form of the constitutive equation of a material is sufficient to describe the material, all forms of the constitutive equation of the material are needed in order to prove that its formulation is thermodynamically consistent.

2.1 A short overview of the nonlinear continuum mechanical theory for finite deformations of elastoplastic media

The nonlinear continuum theory for finite deformations of elastic media has been developed in an elegant manner in the past decades [15–19]. The theory is particularly suitable for modelling elastic materials, whose constitutive equations are defined either in terms of a finite-strain tensor, as in the case of the St-Venant-Kirchhoff material, or derived from an appropriate strain energy density function, as in the case of the hyperelastic materials [17, 18]. Developing material models for finite-strain elastoplasticity within the framework of thermodynamics with internal variables of state, however, requires a somewhat different approach [20]. The constitutive and evolution equations of these materials either exist in rate forms only, or contain rate equations, which at some point during the solution process have to be integrated. Moreover, in nonlinear continuum mechanics, no kinematics of motion can be described without the mathematical definitions of the motion, the Lagrangian and Eulerian displacement fields and the deformation gradient, respectively.

Starting with the definitions (see **Figure 1**), the motion $\mathbf{x} = \mathbf{\Phi}(\mathbf{X}, t)$ from the mathematical pint of view is a vector function or vector field, which maps each material point ${}^{0}P \in {}^{0}\Omega$ with a position vector $\mathbf{X}$ in the initial configuration of the body, into a spatial pint $P \in {}^{t}\Omega$ with a position vector $\mathbf{x}$ in the current configuration of the body. The function must exist whenever the body moves, and it determines the position vector of a material particle at each time instant $t \geq 0$ [16]:

$$\left\{\mathbf{x} | \mathbf{x} = \mathbf{\Phi}(\mathbf{X}, t), \text{for } \mathbf{X} \in {}^{0}\Omega, \mathbf{x} \in {}^{t}\Omega \text{ and } t \geq 0\right\}. \tag{1}$$

In Eq. (1) ${}^{0}\Omega$ is the domain of the function, which stands for the volume of the body in its initial configuration; ${}^{t}\Omega$ is the range of the function, which stands for the volume of the body in its current configuration; and t is time. The vector field that connects the position vectors of the material particle is the displacement field. It

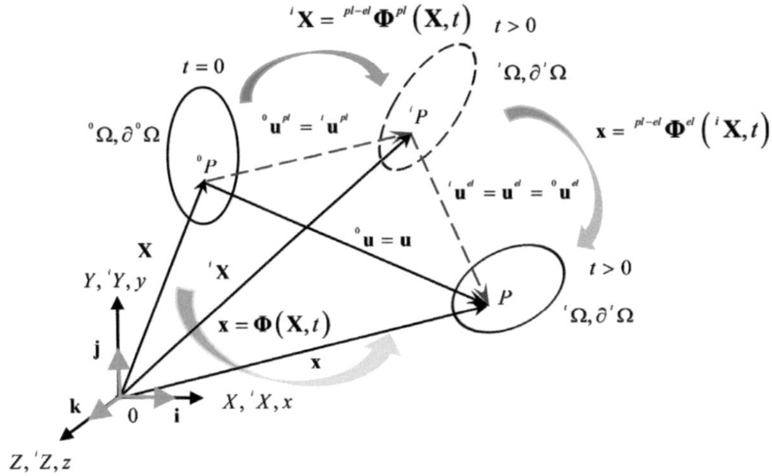

Figure 1.
The proper kinematics of motion of elastoplastic media.

must have both, Lagrangian ${}^{0}\mathbf{u} = {}^{0}\mathbf{u}(\mathbf{X},t)$ and Eulerian $\mathbf{u} = \mathbf{u}(\mathbf{x},t)$ forms; otherwise a physical phenomenon expressed in Lagrangian form cannot be re-expressed in Eulerian form and vice versa. The Lagrangian and Eulerian displacement fields are then defined as [16].

$$^{0}\mathbf{u} = {}^{0}\mathbf{u}(\mathbf{X},t) = \mathbf{x} - \mathbf{X} = \boldsymbol{\Phi}(\mathbf{X},t) - \mathbf{X}, \text{ for } t \geq 0 \text{ and } \mathbf{X} \in {}^{0}\Omega, \quad (2)$$

$$\mathbf{u} = \mathbf{u}(\mathbf{x},t) = \mathbf{x} - \mathbf{X} = \mathbf{x} - \boldsymbol{\Phi}^{-1}(\mathbf{x},t), \text{ for } t \geq 0 \text{ and } \mathbf{x} \in {}^{t}\Omega. \quad (3)$$

The deformation gradient $\mathbf{F}$ then becomes the derivative of the position vector of the material point after motion with respect to the position vector of the material point before motion $\mathbf{F} = \partial\mathbf{x}/\partial\mathbf{X}$ or simply the gradient of the vector function describing the motion.

When the motion is decomposed into several parts, so that the body moves from its initial configuration into its current configuration through several intermediate configurations, the above definitions apply between any two configurations of the body. Let us now consider a deformation process during which the body first undergoes plastic deformations and then elastic deformations at its each constituent. Then the Lagrangian plastic motion ${}^{pl-el}\boldsymbol{\Phi}^{pl} = {}^{pl-el}\boldsymbol{\Phi}^{pl}(\mathbf{X},t)$ and the Lagrangian plastic displacement field ${}^{0}\mathbf{u}^{pl} = {}^{0}\mathbf{u}^{pl}(\mathbf{X},t)$, defined over the initial volume of the body, take the following forms:

$$\left\{ {}^{i}\mathbf{X} \middle| {}^{i}\mathbf{X} = {}^{pl-el}\boldsymbol{\Phi}^{pl}(\mathbf{X},t), \text{for } \mathbf{X} \in {}^{0}\Omega, {}^{i}\mathbf{X} \in {}^{i}\Omega \text{ and } t \geq 0 \right\}, \quad (4)$$

$$^{0}\mathbf{u}^{pl} = {}^{0}\mathbf{u}^{pl}(\mathbf{X},t) = {}^{i}\mathbf{X} - \mathbf{X} = {}^{pl-el}\boldsymbol{\Phi}^{pl}(\mathbf{X},t) - \mathbf{X}, \text{ for } t \geq 0, \text{ and } \mathbf{X} \in {}^{0}\Omega. \quad (5)$$

In Eqs. (4) and (5) ${}^{i}\mathbf{X}$ stands for the position vector of the spatial point ${}^{i}P \in {}^{i}\Omega$, at which the material particle is located when the body has undergone plastic deformations only, and the left superscript ${}^{pl-el}(\bullet)$ denotes the order of elastic and plastic deformations. Although the Eulerian elastic motion ${}^{pl-el}\boldsymbol{\Phi}^{el} = {}^{pl-el}\boldsymbol{\Phi}^{el}({}^{i}\mathbf{X},t)$ and the Eulerian elastic displacement field ${}^{i}\mathbf{u}^{el} = {}^{i}\mathbf{u}^{el}({}^{i}\mathbf{X},t)$ defined over the intermediate volume of the body ${}^{i}\Omega$ have similar forms to the fields above, these represent spatial vector fields, because the intermediate configuration of the body changes during the deformation process.

$$\left\{\mathbf{x}\middle|\mathbf{x} = {}^{pl-el}\boldsymbol{\Phi}^{el}\left({}^{i}\mathbf{X},t\right),\ \text{for}\ {}^{i}\mathbf{X}\in{}^{i}\Omega,\ \mathbf{x}\in{}^{t}\Omega,\ t\geq 0\right\}, \tag{6}$$

$${}^{i}\mathbf{u}^{el} = {}^{i}\mathbf{u}^{el}\left({}^{i}\mathbf{X},t\right) = \mathbf{x} - {}^{i}\mathbf{X} = {}^{pl-el}\boldsymbol{\Phi}^{el}\left({}^{i}\mathbf{X},t\right) - {}^{i}\mathbf{X},\ \text{for}\ t\geq 0\ \text{and}\ {}^{i}\mathbf{X}\in{}^{i}\Omega. \tag{7}$$

Moreover, because the plastic motion exists, the vector fields have Lagrangian forms too. Then the Lagrangian elastic motion ${}^{pl-el}\boldsymbol{\Phi}^{el}\left[{}^{pl-el}\boldsymbol{\Phi}^{pl}(\mathbf{X},t),t\right]$ and the Lagrangian elastic displacement field ${}^{0}\mathbf{u}^{el} = {}^{0}\mathbf{u}^{el}(\mathbf{X},t)$, defined over the initial volume of the body, can be expressed as follows:

$$\left\{\mathbf{x}\middle|\mathbf{x} = {}^{pl-el}\boldsymbol{\Phi}^{el}\left[{}^{pl-el}\boldsymbol{\Phi}^{pl}(\mathbf{X},t),t\right],\ \text{for}\ \mathbf{X}\in{}^{0}\Omega,\ \mathbf{x}\in{}^{t}\Omega,\ t\geq 0\right\}, \tag{8}$$

$${}^{0}\mathbf{u}^{el} = {}^{0}\mathbf{u}^{el}(\mathbf{X},t) = \mathbf{x} - {}^{i}\mathbf{X} = {}^{pl-el}\boldsymbol{\Phi}^{el}\left[{}^{pl-el}\boldsymbol{\Phi}^{pl}(\mathbf{X},t),t\right] - {}^{pl-el}\boldsymbol{\Phi}^{pl}(\mathbf{X},t),\ \text{for}\ t\geq 0\ \text{and}\ \mathbf{X}\in{}^{0}\Omega. \tag{9}$$

Eqs. (1) and (8) then imply the following composite function for the overall motion:

$$\mathbf{x} = \boldsymbol{\Phi}(\mathbf{X},t) = {}^{pl-el}\boldsymbol{\Phi}^{el}\left[{}^{pl-el}\boldsymbol{\Phi}^{pl}(\mathbf{X},t),t\right]. \tag{10}$$

Moreover, after adding Eqs. (5) and (9) up, the following formula for the overall Lagrangian displacement field can be arrived at

$${}^{0}\mathbf{u}^{el}(\mathbf{X},t) + {}^{0}\mathbf{u}^{pl}(\mathbf{X},t) = \boldsymbol{\Phi}^{el}\left[\boldsymbol{\Phi}^{pl}(\mathbf{X},t),t\right] - \mathbf{X} = \mathbf{x} - \mathbf{X} = {}^{0}\mathbf{u}(\mathbf{X},t). \tag{11}$$

Eq. (11) states that the Lagrangian displacement field can additively be decomposed into a Lagrangian elastic part and a Lagrangian plastic part when the kinematics of motion is considered in accordance with the theory of nonlinear continuum mechanics. The deformation gradient then takes the form

$$\mathbf{F} = \mathbf{F}(\mathbf{X},t) = \mathbf{I} + \frac{\partial^{0}\mathbf{u}}{\partial\mathbf{X}} = \mathbf{I} + \frac{\partial^{0}\mathbf{u}^{pl}}{\partial\mathbf{X}} + \frac{\partial^{0}\mathbf{u}^{el}}{\partial\mathbf{X}}. \tag{12}$$

It should be noted that Eq. (12) is the simplest form of the deformation gradient, irrespective of whether the additive decomposition of the displacement field in the above or the multiplicative decomposition of the deformation gradient tensor is used as a starting point in its formulation. In the latter case the formulation would modify as follows:

$$\mathbf{F}(\mathbf{X},t) = \frac{\partial\mathbf{x}}{\partial\mathbf{X}} = \frac{\partial\mathbf{x}}{\partial^{i}\mathbf{X}}\cdot\frac{\partial^{i}\mathbf{X}}{\partial\mathbf{X}} = {}^{pl-el}\mathbf{F}^{el}(\mathbf{X},t)\cdot{}^{pl-el}\mathbf{F}^{pl}(\mathbf{X},t) = \mathbf{I} + \frac{\partial^{0}\mathbf{u}^{el}}{\partial\mathbf{X}} + \frac{\partial^{0}\mathbf{u}^{pl}}{\partial\mathbf{X}}, \tag{13}$$

where

$${}^{pl-el}\mathbf{F}^{el}(\mathbf{X},t) = \frac{\partial\mathbf{x}}{\partial^{i}\mathbf{X}} = \mathbf{I} + \frac{\partial^{i}\mathbf{u}^{el}}{\partial^{i}\mathbf{X}} = \mathbf{I} + \frac{\partial^{0}\mathbf{u}^{el}}{\partial\mathbf{X}}\cdot\left[{}^{pl-el}\mathbf{F}^{pl}(\mathbf{X},t)\right]^{-1}, \tag{14}$$

$$ {}^{pl-el}\mathbf{F}^{pl}(\mathbf{X},t) = \frac{\partial^i \mathbf{X}}{\partial \mathbf{X}} = \mathbf{I} + \frac{\partial^0 \mathbf{u}^{pl}}{\partial \mathbf{X}}, \tag{15} $$

$$ \frac{\partial^i \mathbf{u}^{el}}{\partial^i \mathbf{X}} = \frac{\partial^0 \mathbf{u}^{el}}{\partial \mathbf{X}} \cdot \frac{\partial \mathbf{X}}{\partial^i \mathbf{X}} = \frac{\partial^0 \mathbf{u}^{el}}{\partial \mathbf{X}} \cdot \left[{}^{pl-el}\mathbf{F}^{pl}(\mathbf{X},t) \right]^{-1}, \tag{16} $$

and where considering that ${}^0\mathbf{u}^{el} = {}^i\mathbf{u}^{el}$, in the Lagrangian form Eq. (16) of the Eulerian gradient $\partial^i \mathbf{u}^{el}/\partial^i \mathbf{X}$, the inverse of ${}^{pl-el}\mathbf{F}^{pl}(\mathbf{X},t)$ would be expressed as

$$ \mathbf{I} = \frac{\partial^i \mathbf{X}}{\partial^i \mathbf{X}} = \frac{\partial^i \mathbf{X}}{\partial \mathbf{X}} \cdot \frac{\partial \mathbf{X}}{\partial^i \mathbf{X}} = {}^{pl-el}\mathbf{F}^{pl}(\mathbf{X},t) \cdot \frac{\partial \mathbf{X}}{\partial^i \mathbf{X}} \Leftrightarrow \frac{\partial \mathbf{X}}{\partial^i \mathbf{X}} = \left[{}^{pl-el}\mathbf{F}^{pl}(\mathbf{X},t) \right]^{-1}. \tag{17} $$

Alternatively (see Eq. (10)), the deformation gradient can also be expressed as

$$ \begin{aligned} \mathbf{F}(\mathbf{X},t) &= \frac{\partial \mathbf{x}}{\partial \mathbf{X}} = \frac{\partial^{pl-el}\boldsymbol{\Phi}^{el}\left[{}^{pl-el}\boldsymbol{\Phi}^{pl}(\mathbf{X},t),t \right]}{\partial \mathbf{X}} \\ &= \frac{\partial^{pl-el}\boldsymbol{\Phi}^{el}\left[{}^{pl-el}\boldsymbol{\Phi}^{pl}(\mathbf{X},t),t \right]}{{}^{pl-el}\boldsymbol{\Phi}^{pl}(\mathbf{X},t)} \cdot \frac{{}^{pl-el}\boldsymbol{\Phi}^{pl}(\mathbf{X},t)}{\partial \mathbf{X}} = {}^{pl-el}\mathbf{F}^{el}(\mathbf{X},t) \cdot {}^{pl-el}\mathbf{F}^{pl}(\mathbf{X},t). \end{aligned} \tag{18} $$

It should be noted that by employing the same procedure (Eqs. (13)–(17)), identical formula (Eq. (12)) for the deformation gradient would be arrived at, if the order of elastic and plastic deformations was reversed, although in that case the definitions of the elastic motion ${}^{el-pl}\boldsymbol{\Phi}^{el}(\mathbf{X},t)$, the plastic motion ${}^{el-pl}\boldsymbol{\Phi}^{pl}(\mathbf{X},t)$, the elastic ${}^{el-pl}\mathbf{F}^{el}(\mathbf{X},t)$ and plastic ${}^{el-pl}\mathbf{F}^{pl}(\mathbf{X},t)$ parts of the deformation gradient would be different. The corresponding elastic deformation gradient ${}^{el-pl}\mathbf{F}^{el}(\mathbf{X},t)$, which from now on will be denoted as ${}^{el-pl}\mathbf{F}^{el} = \mathbf{F}^{el}$, then would take the following form:

$$ \mathbf{F}^{el} = \mathbf{I} + \frac{\partial^0 \mathbf{u}^{el}}{\partial \mathbf{X}} = \mathbf{F} - \frac{\partial^0 \mathbf{u}^{pl}}{\partial \mathbf{X}}. \tag{19} $$

When the deformation gradient is in the form of Eq. (12), the material $\dot{\mathbf{E}}$ and the spatial $\mathbf{d} = \mathfrak{L}_e(\mathbf{e})$ strain-rate tensors take the forms

$$ \dot{\mathbf{E}} = \frac{1}{2} \cdot \left(\dot{\mathbf{F}}^T \cdot \mathbf{F} + \mathbf{F}^T \cdot \dot{\mathbf{F}} \right) = \dot{\mathbf{E}}^{el} + \dot{\mathbf{E}}^{pl}, \Rightarrow \mathbf{d} = \mathbf{d}^{el} + \mathbf{d}^{pl}, \tag{20} $$

where

$$ \mathbf{d} = \mathbf{F}^{-T} \cdot \dot{\mathbf{E}} \cdot \mathbf{F}^{-1}, \ \mathbf{d}^{el} = \mathbf{F}^{-T} \cdot \dot{\mathbf{E}}^{el} \cdot \mathbf{F}^{-1}, \ \mathbf{d}^{pl} = \mathbf{F}^{-T} \cdot \dot{\mathbf{E}}^{pl} \cdot \mathbf{F}^{-1}, \tag{21} $$

$$ \dot{\mathbf{E}}^{el} = \frac{1}{2} \cdot \left[\left(\frac{\partial^0 \dot{\mathbf{u}}^{el}}{\partial \mathbf{X}} \right)^T \cdot \mathbf{F} + \mathbf{F}^T \cdot \frac{\partial^0 \dot{\mathbf{u}}^{el}}{\partial \mathbf{X}} \right], \quad \dot{\mathbf{E}}^{pl} = \frac{\dot{\lambda}}{2} \cdot \left[\left(\frac{\partial^P \Psi}{\partial \mathbf{P}} \right)^T \cdot \mathbf{F} + \mathbf{F}^T \cdot \frac{\partial^P \Psi}{\partial \mathbf{P}} \right], \tag{22} $$

$$ \frac{\partial^0 \dot{\mathbf{u}}^{pl}}{\partial \mathbf{X}} = \dot{\lambda} \cdot \frac{\partial^P \Psi}{\partial \mathbf{P}}, \quad \text{and} \quad \frac{\partial^P \Psi}{\partial \mathbf{P}} \neq \left(\frac{\partial^P \Psi}{\partial \mathbf{P}} \right)^T. \tag{23} $$

In the above $\mathbf{E} = 1/2 \cdot (\mathbf{F}^T \cdot \mathbf{F} - \mathbf{I})$ denotes the Green-Lagrangian strain tensor and $\mathbf{e} = 1/2 \cdot (\mathbf{I} - \mathbf{F}^{-T} \cdot \mathbf{F}^{-1})$ the Eulerian-Almansi strain tensor, respectively. The symbols $\dot{\mathbf{E}}^{el}$, $\dot{\mathbf{E}}^{pl}/\mathbf{d}^{el}$, $\mathbf{d}^{pl}$ stand for the elastic and the plastic material/spatial strain-rate tensors, wherein the latter plastic flow is defined by Eq. $(23)_1$ as a product of a plastic multiplier $\dot{\lambda}$ and an appropriate yield surface normal, $\partial^P\Psi/\partial\mathbf{P}$, defined in terms of a first Piola-Kirchhoff stress tensor $\mathbf{P}$. Here the symbol $\mathfrak{L}_e(\mathbf{e}) = \mathbf{F}^{-T} \cdot [\partial(\mathbf{F}^T \cdot (\mathbf{e}) \cdot \mathbf{F})/\partial t] \cdot \mathbf{F}^{-1}$ denotes the Lie derivative of the Eulerian-Almansi strain tensor $\mathbf{e}$. It should also be noted that both elastic and the plastic strain-rate tensors have forms similar to the strain-rate tensor itself. Besides, it can be shown that the plastic flow defined by Eq. $(23)_1$ is not constrained, resulting in Eqs. $(22)_2$ and $(21)_3$, respectively, being the only non-degenerated forms of the material and spatial plastic strain-rate tensors.

A crucial role in the Lagrangian description plays the invariance of the internal mechanical power. It not only defines conjugate pairs of stress measures and strain or deformation rates, but, being the expression of the conservation of internal mechanical energy (first law of thermodynamics), it plays an inevitable role in making sure that the total Lagrangian description and the updated Lagrangian description are equivalent. As a result, appropriate transformations can be found between various stress measures and strain or deformation rates constituting conjugate pairs, when switching from one stress space in one configuration of the body to the other stress space in the same or any other configuration of the body [15–18]. Unfortunately contemporary continuum theory does not cover materials whose constitutive and evolution equations are defined in rate forms. In order to extend the theory, so that it could cover the materials, Cauchy's stress theorem [16] had to be generalised as follows:

$$\mathfrak{L}_{Tr}^{(n)}[\mathbf{T}(\mathbf{X}, t, \mathbf{N})] = \mathfrak{L}_P^{(n)}[\mathbf{P}(\mathbf{X}, t)] \cdot \mathbf{N} \text{ and } \mathfrak{L}_{tr}^{(n)}[\mathbf{t}(\mathbf{x}, t, \mathbf{n})] = \mathfrak{L}_T^{(n)}[\boldsymbol{\sigma}(\mathbf{x}, t)] \cdot \mathbf{n}, \qquad (24)$$

for all $n = 0, 1, \ldots, n \in N$, where n denotes objective differentiation with respect to time $t \geq 0$ and not an exponent and N the set of natural numbers. In Eq. (24) the variables $\mathfrak{L}_{Tr}^{(n)}[\mathbf{T}(\mathbf{X}, t, \mathbf{N})]$, $\mathfrak{L}_{tr}^{(n)}[\mathbf{t}(\mathbf{x}, t, \mathbf{n})]$ stand for the nth objective derivatives of the surface traction vectors $\mathbf{T} = \mathbf{T}(\mathbf{X}, t, \mathbf{N})$, $\mathbf{t} = \mathbf{t}(\mathbf{x}, t, \mathbf{n})$ in the initial and current configurations of the body, and $\mathbf{N}$, $\mathbf{n}$ are the corresponding unit outwards surface normal vectors. Similarly, the quantities $\mathfrak{L}_P^{(n)}[\mathbf{P}(\mathbf{X}, t)]$, $\mathfrak{L}_T^{(n)}[\boldsymbol{\sigma}(\mathbf{x}, t)]$ denote the nth objective derivatives of the first Piola-Kirchhoff stress tensor $\mathbf{P} = \mathbf{P}(\mathbf{X}, t)$ and the Cauchy stress tensor $\boldsymbol{\sigma} = \boldsymbol{\sigma}(\mathbf{x}, t)$, respectively. Then the requirements of thermodynamic consistency of the Lagrangian formulation are ensured by the following postulates:

Postulate no. 1. The product of a surface traction vector, including all its higher-order objective time derivatives and the surface of an infinitesimal volume element in the initial and current configurations of the body, on which they act, have to be the same during the deformation process, i.e.:

$$\begin{aligned} \mathfrak{L}_{Tr}^{(n)}(\mathbf{T}) \cdot dS_0 &= \mathfrak{L}_P^{(n)}(\mathbf{P}) \cdot \mathbf{N} \cdot dS_0 = \mathfrak{L}_P^{(n)}(\mathbf{P}) \cdot d\mathbf{S}_0 = \mathfrak{L}_{tr}^{(n)}(\mathbf{t}) \cdot ds \\ &= \mathfrak{L}_T^{(n)}(\boldsymbol{\sigma}) \cdot \mathbf{n} \cdot ds = \mathfrak{L}_T^{(n)}(\boldsymbol{\sigma}) \cdot d\mathbf{s}, \end{aligned} \qquad (25)$$

or

$$\mathfrak{L}_P^{(n)}(\mathbf{P}) \cdot d\mathbf{S}_0 = \mathfrak{L}_T^{(n)}(\boldsymbol{\sigma}) \cdot d\mathbf{s} = J \cdot \mathfrak{L}_T^{(n)}(\boldsymbol{\sigma}) \cdot \mathbf{F}^{-T} \cdot d\mathbf{S}_0, \text{ for all } n = 0, 1, 2, \ldots, n \in N, \qquad (26)$$

where $d\mathbf{S}_0$, $d\mathbf{s} = J \cdot \mathbf{F}^{-T} \cdot d\mathbf{S}_0$ denote the infinitesimal surface elements in the initial and current configurations of the body, where the latter is expressed using the Nanson's formula [16].

Postulate no. 2. The rate of change of the internal mechanical energy accumulated in the infinitesimal volume element in the initial and current configurations of the body and all its higher-order time derivatives have to be the same during the deformation process, i.e.:

$$\frac{\partial^n dW}{\partial t^n} = \left[\sum_{k=0}^{n} \binom{n}{k} \cdot \frac{\partial^{n-k}\mathbf{S}}{\partial t^{n-k}} : \frac{\partial^k \dot{\mathbf{E}}}{\partial t^k}\right] \cdot dV_0 = \left[\sum_{k=0}^{n} \binom{n}{k} \cdot \mathfrak{L}_P^{(n-k)}(\mathbf{P}) : \mathfrak{L}_F^{(k)}\left(\frac{\partial^0 \dot{\mathbf{u}}}{\partial \mathbf{X}}\right)\right]$$
$$\cdot dV_0 = \left[\sum_{k=0}^{n} \binom{n}{k} \cdot \mathfrak{L}_O^{(n-k)}(\boldsymbol{\tau}) : \mathfrak{L}_e^{(k)}(\mathbf{d})\right] \cdot dV_0 = \left[\sum_{k=0}^{n} \binom{n}{k} \cdot \mathfrak{L}_T^{(n-k)}(\boldsymbol{\sigma}) : \mathfrak{L}_e^{(k)}(\mathbf{d})\right]$$
$$\cdot dv, \quad \text{for all } n = 0, 1, 2, \ldots, n \in N, \tag{27}$$

or

$$\frac{\partial^n dW}{\partial t^n} = \left[\sum_{k=0}^{n} \binom{n}{k} \cdot \frac{\partial^{n-k}\mathbf{S}}{\partial t^{n-k}} : \frac{\partial^k \dot{\mathbf{E}}}{\partial t^k}\right] \cdot dV_0 = \left[\sum_{k=0}^{n} \binom{n}{k} \cdot \mathbf{F} \cdot \left(\frac{\partial^{n-k}\mathbf{S}}{\partial t^{n-k}}\right) : \mathbf{F}^{-T} \cdot \frac{\partial^k \left(\mathbf{F}^T \cdot \dot{\mathbf{F}}\right)}{\partial t^k}\right]$$
$$\cdot dV_0 = \left[\sum_{k=0}^{n} \binom{n}{k} \cdot \mathbf{F} \cdot \left(\frac{\partial^{n-k}\mathbf{S}}{\partial t^{n-k}}\right) \cdot \mathbf{F}^T : \mathbf{F}^{-T} \cdot \left(\frac{\partial^k \dot{\mathbf{E}}}{\partial t^k}\right) \cdot \mathbf{F}^{-1}\right] \cdot dV_0 =$$
$$= \left[\sum_{k=0}^{n} \binom{n}{k} \cdot \frac{\mathbf{F}}{J} \cdot \left(\frac{\partial^{n-k}\mathbf{S}}{\partial t^{n-k}}\right) \cdot \mathbf{F}^T : \mathbf{F}^{-T} \cdot \left(\frac{\partial^k \dot{\mathbf{E}}}{\partial t^k}\right) \cdot \mathbf{F}^{-1}\right] \cdot dv \quad \text{for all } n = 0, 1, 2, \ldots, n \in N. \tag{28}$$

where $dV_0, dv = J \cdot dV_0$ stand for the infinitesimal volume elements in the initial and current configurations of the body and $J = \det(\mathbf{F})$. Then Eqs. (26)–(28) define the following transformations:

$$\mathfrak{L}_P^{(n)}(\mathbf{P}) = \mathbf{F} \cdot \left(\frac{\partial^n \mathbf{S}}{\partial t^n}\right), \mathfrak{L}_O^{(n)}(\boldsymbol{\tau}) = \mathbf{F} \cdot \left(\frac{\partial^n \mathbf{S}}{\partial t^n}\right) \cdot \mathbf{F}^T, \quad \mathfrak{L}_T^{(n)}(\boldsymbol{\sigma}) = J^{-1} \cdot \mathbf{F} \cdot \left(\frac{\partial^n \mathbf{S}}{\partial t^n}\right) \cdot \mathbf{F}^T,$$
$$\mathfrak{L}_e^{(n)}(\mathbf{d}) = \mathbf{F}^{-T} \cdot \left(\frac{\partial^n \dot{\mathbf{E}}}{\partial t^n}\right) \cdot \mathbf{F}^{-1}, \quad \frac{\partial^n \dot{\mathbf{E}}}{\partial t^n} = \left[\mathbf{F}^T \cdot \mathfrak{L}_F^{(n)}(\dot{\mathbf{F}})\right]^{sym} = \left[\frac{\partial^n \left(\mathbf{F}^T \cdot \frac{\partial^0 \dot{\mathbf{u}}}{\partial \mathbf{X}}\right)}{\partial t^n}\right]^{sym},$$
$$\mathfrak{L}_e^{(n)}(\mathbf{d}) = \left[\mathfrak{L}_F^{(n)}(\dot{\mathbf{F}}) \cdot \mathbf{F}^{-1}\right]^{sym} = \left\{\mathbf{F}^{-T} \cdot \left[\frac{\partial^n \left(\mathbf{F}^T \cdot \frac{\partial^0 \dot{\mathbf{u}}}{\partial \mathbf{X}}\right)}{\partial t^n}\right] \cdot \mathbf{F}^{-1}\right\}^{sym} \quad \text{for all } n = 0, 1, \ldots, n \in N, \tag{29}$$

as the sufficient conditions of thermodynamic consistency, because they ensure that the two postulates above are met. It should also be noted that for $n = 0$ the transformations define the necessary conditions of thermodynamic consistency. In that case the generalised Cauchy's stress theorem Eq. (24) reduces to its well-known form, $\mathbf{T} = \mathbf{P} \cdot \mathbf{N}$ and $\mathbf{t} = \boldsymbol{\sigma} \cdot \mathbf{n}$, , while the transformations Eq. (29) reduce to the already well-known transformations in nonlinear continuum mechanics, defining the relationship between various stress measures and strain or deformation rates constituting the conjugate pairs.

The objective rates, which meet the sufficient conditions of thermodynamic consistency defined by Eq. (29), are already known in nonlinear continuum mechanics as the $\mathrm{n^{th}}$ Lie derivative of the first Piola-Kirchhoff stress tensor $\mathbf{P}$ (Eq. (30)), the $\mathrm{n^{th}}$ Lie derivative of the rate of deformation gradient tensor $\dot{\mathbf{F}}$ (Eq. (31)), the $\mathrm{n^{th}}$ Oldroyd derivative of the Kirchhoff stress $\boldsymbol{\tau}$ tensor (Eq. (32)), the $\mathrm{n^{th}}$ Lie derivative of the spatial strain-rate tensor $\mathbf{d}$ (Eq. (33)) and the $\mathrm{n^{th}}$ Truesdell derivative of the Cauchy stress tensor Eq. (34), respectively:

$$\mathfrak{L}_P^{(n)}(\mathbf{P}) = \mathbf{F} \cdot \left[\frac{\partial^n (\mathbf{F}^{-1} \cdot \mathbf{P})}{\partial t^n} \right] = \mathbf{F} \cdot \left(\frac{\partial^n \mathbf{S}}{\partial t^n} \right), \tag{30}$$

$$\mathfrak{L}_F^{(n)}(\dot{\mathbf{F}}) = \mathbf{F}^{-T} \cdot \left[\frac{\partial^n (\mathbf{F}^T \cdot \dot{\mathbf{F}})}{\partial t^n} \right] = \mathbf{F}^{-T} \cdot \frac{\partial^n \left(\mathbf{F}^T \cdot \frac{\partial^0 \dot{\mathbf{u}}}{\partial \mathbf{X}} \right)}{\partial t^n}, \tag{31}$$

$$\mathfrak{L}_O^{(n)}(\boldsymbol{\tau}) = \mathbf{F} \cdot \left[\frac{\partial^n (\mathbf{F}^{-1} \cdot \boldsymbol{\tau} \cdot \mathbf{F}^{-T})}{\partial t^n} \right] \cdot \mathbf{F}^T = \mathbf{F} \cdot \left(\frac{\partial^n \mathbf{S}}{\partial t^n} \right) \cdot \mathbf{F}^T, \tag{32}$$

$$\mathfrak{L}_e^{(n)}(\mathbf{d}) = \mathbf{F}^{-T} \cdot \left[\frac{\partial^n (\mathbf{F}^T \cdot \mathbf{d} \cdot \mathbf{F})}{\partial t^n} \right] \cdot \mathbf{F}^{-1} = \mathbf{F}^{-T} \cdot \left(\frac{\partial^n \dot{\mathbf{E}}}{\partial t^n} \right) \cdot \mathbf{F}^{-1}, \tag{33}$$

$$\mathfrak{L}_T^{(n)}(\boldsymbol{\sigma}) = J^{-1} \cdot \mathbf{F} \cdot \left[\frac{\partial^n (J \cdot \mathbf{F}^{-1} \cdot \boldsymbol{\sigma} \cdot \mathbf{F}^{-T})}{\partial t^n} \right] \cdot \mathbf{F}^T = J^{-1} \cdot \mathbf{F} \cdot \left(\frac{\partial^n \mathbf{S}}{\partial t^n} \right) \cdot \mathbf{F}^T. \tag{34}$$

It should also be noted here that Eq. (27) is the result of straightforward manipulation of the nth time derivative of the internal mechanical power (see Eq. (28)), whose last terms formally define the formulas for evaluating the nth objective derivative of the power with respect to time in the first Piola-Kirchhoff, Kirchhoff and Cauchy stress spaces. Then Eq. (27) defines not only conjugate pairs of stress measures and strain or deformations rates but also conjugate pairs of objective differentiation operators and derivatives. Moreover, we used intentionally the term 'requirements of thermodynamic consistency' for the transformations Eq. (29), because without the invariance of the internal mechanical power (first law of thermodynamics) and its higher-order time derivatives (Eq. (27) or (28)), no formulation is thermodynamically consistent, in spite of the fact that in thermodynamics the term is associated with the second law of thermodynamics to show that the constitutive equation of a material is compatible with the second law.

2.2 Modelling of the plastic flow in the material

In order to modify the nonlinear continuum theory for finite deformations of elastoplastic media, it is assumed that the yield surface of the material has definitions ${}^S\Psi = {}^S\Psi\left[{}^S\sigma_{eq}(\mathbf{S}), \mathbf{q}\right]$, ${}^P\Psi = {}^P\Psi\left[{}^P\sigma_{eq}(\mathbf{P}), \mathbf{q}\right]$, ${}^\tau\Psi = {}^\tau\Psi\left[{}^\tau\sigma_{eq}(\boldsymbol{\tau}), \mathbf{q}\right]$, ${}^\sigma\Psi = {}^\sigma\Psi \left[{}^\sigma\sigma_{eq}(\boldsymbol{\sigma}), \mathbf{q}\right]$ in terms of the second Piola-Kirchhoff stress tensor $\mathbf{S}$, the first Piola-Kirchhoff stress tensor $\mathbf{P}$, the Kirchhoff stress tensor $\boldsymbol{\tau}$, the Cauchy stress tensor $\boldsymbol{\sigma}$ and a vector of internal variables $\mathbf{q}$ in the second Piola-Kirchhoff, first Piola-Kirchhoff, Kirchhoff and Cauchy stress spaces, where ${}^S\sigma_{eq}(\mathbf{S})$, ${}^P\sigma_{eq}(\mathbf{P})$, ${}^\sigma\sigma_{eq}(\boldsymbol{\tau})$, ${}^\sigma\sigma_{eq}(\boldsymbol{\sigma})$ are the corresponding equivalent stresses. After changing the physical interpretation of the plastic flow and applying push-forward and pull-back operations to the material gradient of the plastic velocity field, Eq. $(23)_1$ is as follows:

$$\frac{\partial \dot{\mathbf{u}}^{pl}}{\partial \mathbf{x}} = \frac{\partial^0 \dot{\mathbf{u}}^{pl}}{\partial \mathbf{X}} \cdot \mathbf{F}^{-1} = \dot{\lambda} \cdot \frac{\partial^\sigma \Psi}{\partial \boldsymbol{\sigma}}, \ \frac{\partial \dot{\mathbf{u}}^{pl}}{\partial \mathbf{x}} = \frac{\partial^0 \dot{\mathbf{u}}^{pl}}{\partial \mathbf{X}} \cdot \mathbf{F}^{-1} = \dot{\lambda} \cdot \frac{\partial^\tau \Psi}{\partial \boldsymbol{\tau}}, \mathbf{F}^T \cdot \frac{\partial^0 \dot{\mathbf{u}}^{pl}}{\partial \mathbf{X}} = \dot{\lambda} \cdot \frac{\partial^S \Psi}{\partial \mathbf{S}}, \tag{35}$$

where it can be shown that the definitions of the yield surface are not independent of each other but are related, and the following formulas hold true:

$$\frac{\partial^P \Psi}{\partial \mathbf{P}} \cdot \mathbf{F}^{-1} = \frac{\partial^\sigma \Psi}{\partial \boldsymbol{\sigma}}, \quad \frac{\partial^P \Psi}{\partial \mathbf{P}} \cdot \mathbf{F}^{-1} = \frac{\partial^\tau \Psi}{\partial \boldsymbol{\tau}}, \quad \mathbf{F}^T \cdot \frac{\partial^P \Psi}{\partial \mathbf{P}} = \frac{\partial^S \Psi}{\partial \mathbf{S}}. \tag{36}$$

As a result, one of the definitions of the yield surfaces has to be chosen as a reference to define the material model, and the rest of them can be calculated by solving the differential equations in Eq. (36). Moreover, when ${}^\sigma\Psi$ or ${}^\tau\Psi$ is used as the reference definition of the yield surface in the current configuration of the body, the contemporary flow plasticity models will be recovered. It also can be verified that the various definitions of the yield surface and their equivalent stresses ${}^P\sigma_{eq}, {}^S\sigma_{eq}, {}^\tau\sigma_{eq}, {}^\sigma\sigma_{eq}$, which also meet the transformations defined by Eq. (36), have the following properties:

$${}^P\sigma_{eq} = {}^S\sigma_{eq} = {}^\tau\sigma_{eq} = J \cdot {}^\sigma\sigma_{eq}, \tag{37}$$

$$\frac{\partial^S \Psi}{\partial \mathbf{S}} : \mathbf{S} = \frac{\partial^P \Psi}{\partial \mathbf{P}} : \mathbf{P} = \frac{\partial^\tau \Psi}{\partial \boldsymbol{\tau}} : \boldsymbol{\tau} = J \cdot \frac{\partial^\sigma \Psi}{\partial \boldsymbol{\sigma}} : \boldsymbol{\sigma}, \tag{38}$$

$$\frac{\partial^S \Psi}{\partial \mathbf{S}} : \dot{\mathbf{S}} = \frac{\partial^P \Psi}{\partial \mathbf{P}} : \mathfrak{L}_P(\mathbf{P}) = \frac{\partial^\tau \Psi}{\partial \boldsymbol{\tau}} : \mathfrak{L}_O(\boldsymbol{\tau}) = J \cdot \frac{\partial^\sigma \Psi}{\partial \boldsymbol{\sigma}} : \mathfrak{L}_T(\boldsymbol{\sigma}), \tag{39}$$

where Eqs. (38) and (39) represent 'normality rules', which from the physical point of view are equivalent with the following equations:

$$dW^{pl} = \dot{\mathbf{E}}^{pl} : \mathbf{S} \cdot dV_0 = \frac{\partial^0 \dot{\mathbf{u}}^{pl}}{\partial \mathbf{X}} : \mathbf{P} \cdot dV_0 = \mathbf{d}^{pl} : \boldsymbol{\tau} \cdot dV_0 = \mathbf{d}^{pl} : \boldsymbol{\sigma} \cdot dv, \tag{40}$$

$$\dot{\mathbf{E}}^{pl} : \dot{\mathbf{S}} \cdot dV_0 = \frac{\partial^0 \dot{\mathbf{u}}^{pl}}{\partial \mathbf{X}} : \mathfrak{L}_P(\mathbf{P}) \cdot dV_0 = \mathbf{d}^{pl} : \mathfrak{L}_O(\boldsymbol{\tau}) \cdot dV_0 = \mathbf{d}^{pl} : \mathfrak{L}_T(\boldsymbol{\sigma}) \cdot dv. \tag{41}$$

where W^{pl} is the internal plastic power.

2.3 The constitutive equations of the material

Proper formulation of a material model for finite-strain elastoplasticity allows for the definition of the model in all stress spaces in any configuration of the body. These, however, have to comply with the principles of material modelling, particularly to meet the requirements of material objectivity and be thermodynamically consistent in order that they would define the same material. Finite-strain computational plasticity distinguishes between two major types of material models known as hypoelastic-plastic-based material models and hyperelastic-plastic-based material models. Moreover, hypoelastic-plastic-based material models exist in rate forms only, because the additive decomposition of the strain-rate tensor Eqs. (20)–(23) and (36) exists either in rate forms only. In this research we have modified our former material model with internal damping, capable of imitating even ductile-to-brittle failure mode transition at high strain rates, to model our hypoelastic-plastic-based material [21]. The rate form of the constitutive equation of the material then can take any of the following forms:

$$\dot{\mathbf{S}} = {}^{mat}\mathbb{C}^{el} : \left(\dot{\mathbf{E}} - xx \cdot \dot{\mathbf{E}}^{pl}\right) + {}^{mat}\mathbb{C}^{vis} : \left[\ddot{\mathbf{E}} - (1 - xx) \cdot \ddot{\mathbf{E}}^{pl}\right], \tag{42}$$

$$\mathfrak{L}_P(\mathbf{P}) = \mathbf{F}\cdot\dot{\mathbf{S}} = {}^{mix}\mathbb{C}^{el} : \left(\dot{\mathbf{F}} - xx\cdot\frac{\partial^0 \dot{\mathbf{u}}^{pl}}{\partial \mathbf{X}}\right) + {}^{mix}\mathbb{C}^{vis} : \left[\mathfrak{L}_F(\dot{\mathbf{F}}) - (1-xx)\cdot\mathfrak{L}_F\left(\frac{\partial^0 \dot{\mathbf{u}}^{pl}}{\partial \mathbf{X}}\right)\right], \tag{43}$$

$$\mathfrak{L}_O(\boldsymbol{\tau}) = \mathbf{F}\cdot\dot{\mathbf{S}}\cdot\mathbf{F}^T = J\cdot{}^{spat}\mathbb{C}^{el} : \left(\mathbf{d} - xx\cdot\mathbf{d}^{pl}\right) + J\cdot{}^{spat}\mathbb{C}^{vis} : \left[\mathfrak{L}_e(\mathbf{d}) - (1-xx)\cdot\mathfrak{L}_e\left(\mathbf{d}^{pl}\right)\right], \tag{44}$$

$$\mathfrak{L}_T(\boldsymbol{\sigma}) = J^{-1}\cdot\mathbf{F}\cdot\dot{\mathbf{S}}\cdot\mathbf{F}^T = {}^{spat}\mathbb{C}^{el} : \left(\mathbf{d} - xx\cdot\mathbf{d}^{pl}\right) + {}^{spat}\mathbb{C}^{vis} : \left[\mathfrak{L}_e(\mathbf{d}) - (1-xx)\cdot\mathfrak{L}_e\left(\mathbf{d}^{pl}\right)\right], \tag{45}$$

where

$${}^{mat}\mathbb{C}^{el} = 2\cdot G\cdot\mathbb{I} + \lambda^{el}\cdot\mathbf{I}\otimes\mathbf{I}, \quad {}^{mat}\mathbb{C}^{vis} = 2\cdot G^{vis}\cdot\mathbb{I} + \lambda^{vis}\cdot\mathbf{I}\otimes\mathbf{I}, \tag{46}$$

$$G = \frac{E}{2\cdot(1+\nu)}, \lambda^{el} = \frac{\nu\cdot E}{(1+\nu)\cdot(1-2\cdot\nu)}, \tag{47}$$

$$G^{vis} = \frac{E^{vis}}{2\cdot(1+\nu^{vis})}, \lambda^{vis} = \frac{\nu^{vis}\cdot E^{vis}}{(1+\nu^{vis})\cdot(1-2\cdot\nu^{vis})}, \tag{48}$$

$${}^{mix}\mathbb{C}^{el}_{ijkl} = F_{im}\cdot\delta_{jn}\cdot F_{ko}\cdot\delta_{lp}\cdot{}^{mat}\mathbb{C}^{el}_{mnop}, \quad {}^{mix}\mathbb{C}^{vis}_{ijkl} = F_{im}\cdot\delta_{jn}\cdot F_{ko}\cdot\delta_{lp}\cdot{}^{mat}\mathbb{C}^{vis}_{mnop}, \tag{49}$$

$${}^{spat}\mathbb{C}^{el}_{ijkl} = J^{-1}\cdot F_{im}\cdot F_{jn}\cdot F_{ko}\cdot F_{lp}\cdot{}^{mat}\mathbb{C}^{el}_{mnop}, \quad {}^{spat}\mathbb{C}^{vis}_{ijkl} = J^{-1}\cdot F_{im}\cdot F_{jn}\cdot F_{ko}\cdot F_{lp}\cdot{}^{mat}\mathbb{C}^{vis}_{mnop}. \tag{50}$$

In Eqs. (42)–(50) the symbols $\dot{\mathbf{S}}$, $\mathfrak{L}_P(\mathbf{P})$, $\mathfrak{L}_O(\boldsymbol{\tau})$, $\mathfrak{L}_T(\boldsymbol{\sigma})$ denote the time derivative of the second Piola-Kirchhoff stress tensor, the Lie derivative of the first Piola-Kirchhoff stress tensor, the Oldroyd rate of the Kirchhoff stress and the Truesdell rate of the Cauchy stress, respectively. Here the fourth-order material elasticity tensor ${}^{mat}\mathbb{C}^{el}$ and the fourth-order material viscosity tensor ${}^{mat}\mathbb{C}^{vis}$ have similar forms as the fourth-order elasticity tensor of the St. Venant-Kirchhoff material [17] using two independent material parameters E, ν and E^{vis}, ν^{vis}, respectively, where $\mathbb{I}$ denotes the symmetric fourth-order identity tensor and $\mathbf{I}$ the second-order identity tensor. The fourth-order mixed spatial-material elasticity and viscosity tensors ${}^{mat}\mathbb{C}^{el}$, ${}^{mat}\mathbb{C}^{vis}$ then can be determined in accordance with Eq. (49), where δ_{ij} is the Kronecker delta, and the fourth-order spatial elasticity and viscosity tensors in accordance with Eq. (50). The variable xx denotes the ratio of ductile and total damage increment [21]. It should be noted that the objective rates $\dot{\mathbf{S}}$, $\mathfrak{L}_P(\mathbf{P})$, $\mathfrak{L}_O(\boldsymbol{\tau})$, $\mathfrak{L}_T(\boldsymbol{\sigma})$ transform in the same way from one form into another as do the stress tensors $\mathbf{S}$, $\mathbf{P}$, $\boldsymbol{\tau}$, $\boldsymbol{\sigma}$, which ensure that the formulation is thermodynamically consistent (see also Eq. (29)). Then, the corresponding rate forms of loading/unloading discrete Kuhn-Tucker plastic optimization conditions in the second Piola-Kirchhoff (Eq. (51)), first Piola-Kirchhoff (Eq. (52)), Kirchhoff (Eq. (53)) and Cauchy (Eq. (54)) stress spaces are modified as follows:

$$\dot{\lambda} \geq 0, \qquad {}^{S}\Psi \leq 0, \qquad \dot{\lambda} \cdot {}^{S}\dot{\Psi} = 0, \tag{51}$$

$$\dot{\lambda} \geq 0, \qquad {}^{P}\Psi \leq 0, \qquad \dot{\lambda} \cdot \mathfrak{L}_P\left({}^{P}\Psi\right) = 0, \tag{52}$$

$$\dot{\lambda} \geq 0, \qquad {}^{\tau}\Psi \leq 0, \qquad \dot{\lambda} \cdot \mathfrak{L}_O\left({}^{\tau}\Psi\right) = 0, \tag{53}$$

$$\dot{\lambda} \geq 0, \qquad {}^{\sigma}\Psi \leq 0, \qquad \dot{\lambda} \cdot \mathfrak{L}_T\left({}^{\sigma}\Psi\right) = 0. \tag{54}$$

It should be noted here that the above conditions resulted from the invariance of the internal mechanical power and its first time derivative (see Eq. (27)), which now also define conjugate pairs of differentiation operators and derivatives in all stress spaces, and that all yield surface is the expression of the conservation of internal plastic power, which will be shown later.

Hyperelastic-plastic-based material models are essentially elastic material models. The starting point in their development is Eq. (19), wherein the constitutive equation of the material the elastic Green-Lagrangian strain tensor and its time derivative are modified as follows:

$$^{*}\mathbf{E}^{el} = \frac{1}{2} \cdot \left[\left(\mathbf{F}^{el}\right)^{T} \cdot \mathbf{F}^{el} - \mathbf{I}\right], \qquad ^{*}\dot{\mathbf{E}}^{el} = \frac{1}{2} \cdot \left[\left(\dot{\mathbf{F}}^{el}\right)^{T} \cdot \mathbf{F}^{el} + \left(\mathbf{F}^{el}\right)^{T} \cdot \dot{\mathbf{F}}^{el}\right]. \tag{55}$$

In our research, the St. Venant hyperelastic material with internal damping was used, whose constitutive equation then takes any of the following forms:

$$\mathbf{S} = {}^{mat}\mathbb{C}^{el} : {}^{*}\mathbf{E}^{el} + {}^{mat}\mathbb{C}^{vis} : {}^{*}\dot{\mathbf{E}}^{el}, \tag{56}$$

$$\mathbf{P} = \mathbf{F} \cdot \mathbf{S}, \qquad \boldsymbol{\tau} = \mathbf{F} \cdot \mathbf{S} \cdot \mathbf{F}^{T}, \qquad \boldsymbol{\sigma} = J^{-1} \cdot \mathbf{F} \cdot \mathbf{S} \cdot \mathbf{F}^{T}. \tag{57}$$

In the above equations and also in (19), the incremental form of the material gradient of the plastic displacement field is determined as follows:

$$\frac{\partial^{n+1,0}\mathbf{u}^{pl}}{\partial \mathbf{X}} = \Delta t \cdot \frac{\partial^{0}\dot{\mathbf{u}}^{pl}}{\partial \mathbf{X}} + \frac{\partial^{n,0}\mathbf{u}^{pl}}{\partial \mathbf{X}}, \tag{58}$$

where the material gradient of the plastic velocity field $\partial^0 \dot{\mathbf{u}} pl/\partial \mathbf{X}$ is calculated in accordance with Eq. $(23)_1$ or Eq. (36), depending on the reference definition of the yield surface. The corresponding loading/unloading discrete Kuhn-Tucker plastic optimization conditions then take the forms:

$$\dot{\lambda} \geq 0, \qquad {}^{S}\Psi \leq 0, \qquad \dot{\lambda} \cdot {}^{S}\Psi = 0, \tag{59}$$

$$\dot{\lambda} \geq 0, \qquad {}^{P}\Psi \leq 0, \qquad \dot{\lambda} \cdot {}^{P}\Psi = 0, \tag{60}$$

$$\dot{\lambda} \geq 0, \qquad {}^{\tau}\Psi \leq 0, \qquad \dot{\lambda} \cdot {}^{\tau}\Psi = 0, \tag{61}$$

$$\dot{\lambda} \geq 0, \qquad {}^{\sigma}\Psi \leq 0, \qquad \dot{\lambda} \cdot {}^{\sigma}\Psi = 0. \tag{62}$$

2.4 The reference definition of the yield surface

Objective and thermodynamically consistent formulation of the plastic flow allows for the development of consistent material models. As a result, the material model can by formulated in any stress space and in whatever configuration of the body, though we intentionally omitted to give an example of the constitutive equation of our hypoelastic-plastic-based and hyperelastic-plastic-based material models in the intermediate configuration of the body, as it is just a matter of proper stress transformation using the multiplicative split of the deformation. Moreover, one of the formulations of the yield surface has to be a reference, from which other definitions of the yield surface in the second Piola-Kirchhoff, first Piola-Kirchhoff, Kirchhoff and Cauchy stress spaces can be calculated by solving Eq. (36). The various definitions of the yields surface then have the properties Eqs. (37)–(39), from which the 'normality rules' (Eqs. (38) and (39)) (whose physical meaning is defined by Eqs. (40) and (41)) are used in the return mapping/rate form of the return mapping algorithms to calculate the plastic multiplier. In this study the reference yield surface was defined in the first Piola-Kirchhoff stress space, because the corresponding plastic flow Eq. $(23)_1$ is the only not constrained. Then the generalised J_2 flow plasticity theory with isotropic hardening is defined by Eqs. (63)–(67). It should be noted that the ${}^{P}J_2(\mathbf{P}) = \mathbf{P} : \mathbf{P}$ invariant in the definition of the equivalent stress no longer bases on the deviatoric part of the first Piola-Kirchhoff stress tensor. The change was implied by the objectivity requirements, since the first Piola-Kirchhoff stress tensor transforms under the change of the observer as $\mathbf{P}^+ = \mathbf{Q}_R \cdot \mathbf{P}$ and ${}^{P}J_2(\mathbf{P})$ is the only invariant, which is not affected by the change, i.e. ${}^{P}J_2(\mathbf{P}) = {}^{P}J_2(\mathbf{P}^+)$, where $\mathbf{Q}_R$ is a rotating tensor expressing the relative rotation of the coordinate systems of an arbitrarily moving observer with respect to the reference coordinate system. The resulting yield surface is then no longer a cylinder but a sphere:

$$ {}^{P}\Psi = {}^{P}\sigma_{eq} - {}^{P}\sigma_y \leq 0, \quad \text{where} \quad {}^{P}\sigma_{eq} = {}^{P}\sigma_{eq}(\mathbf{P}) = \sqrt{{}^{P}J_2(\mathbf{P})} = \sqrt{\mathbf{P} : \mathbf{P}}, \tag{63} $$

$$ {}^{P}\sigma_y = F_{UT11} \cdot \sqrt{r^2 - \left[a \cdot e^{pl} - \text{center}\right]^2}, r = \sigma_y + Q, \ \text{center} = \sqrt{r^2 - \sigma_y^2} \ \text{and} \ a = \frac{\text{center} + r}{b}, \tag{64} $$

$$ \begin{aligned} {}^{spring}\dot{e}^{pl} &= {}^{spring}\dot{e}^{pl}\left(xx \cdot \frac{\partial^0 \dot{\mathbf{u}}^{pl}}{\partial \mathbf{X}}\right) = \sqrt{xx \cdot \frac{\partial^0 \dot{\mathbf{u}}^{pl}}{\partial \mathbf{X}} : xx \cdot \frac{\partial^0 \dot{\mathbf{u}}^{pl}}{\partial \mathbf{X}}} = xx \cdot \dot{\lambda}, \\ {}^{spring}e^{pl} &= \int_0^t {}^{spring}\dot{e}^{pl} \cdot dt = \int_0^t xx \cdot \dot{\lambda} \cdot dt, \qquad \text{with} \qquad xx \in \langle 0; 1 \rangle, \end{aligned} \tag{65} $$

$$ \begin{aligned} {}^{damper}\dot{e}^{pl} &= {}^{damper}\dot{e}^{pl}\left((1-xx) \cdot \frac{\partial^0 \dot{\mathbf{u}}^{pl}}{\partial \mathbf{X}}\right) = \sqrt{(1-xx) \cdot \frac{\partial^0 \dot{\mathbf{u}}^{pl}}{\partial \mathbf{X}} : (1-xx) \cdot \frac{\partial^0 \dot{\mathbf{u}}^{pl}}{\partial \mathbf{X}}} = (1-xx) \cdot \dot{\lambda}, \\ {}^{damper}e^{pl} &= \int_0^t {}^{damper}\dot{e}^{pl} \cdot dt = \int_0^t (1-xx) \cdot \dot{\lambda} \cdot dt, \qquad \text{with} \qquad xx \in \langle 0; 1 \rangle, \end{aligned} \tag{66} $$

$$\dot{e}^{pl} = \dot{e}^{pl}\left(\dot{\mathbf{F}}^{pl}\right) = \sqrt{\frac{\partial^0 \dot{\mathbf{u}}^{pl}}{\partial \mathbf{X}} : \frac{\partial^0 \dot{\mathbf{u}}^{pl}}{\partial \mathbf{X}}} = \dot{\lambda}, \quad e^{pl} = \int_0^t \dot{e}^{pl} \cdot dt = \int_0^t \dot{\lambda} \cdot dt, \quad \frac{\partial^0 \dot{\mathbf{u}}^{pl}}{\partial \mathbf{X}} = \dot{\lambda} \cdot \frac{\partial^P \Psi}{\partial \mathbf{X}}. \tag{67}$$

The actual yield stress ${}^P\sigma_y$, which is a first Piola-Kirchhoff stress measure, determines the radius of the yield surface and is defined by Eq. $(64)_1$. It is the only nonzero component of a stress tensor $\mathbf{P}_{UT}$ (i.e. ${}^P\sigma_y = \mathrm{P}_{UT11} = [\mathbf{P}_{UT}]_{11}$) coming from a uniaxial tensile test of the modelled material, where the operator $[(\bullet)]_{11}$ extracts the element in the first row and the first column of a second-order tensor, $(\bullet)$, is written as a 3×3 square matrix. The corresponding deformation gradient and the Jacobian of deformation are denoted as $\mathbf{F}_{UT}, J_{UT}$, where $F_{UT11} = [\mathbf{F}_{UT}]_{11}$ and $J_{UT} = \det(\mathbf{F}_{UT})$. Please also note that the only nonzero element of the related second Piola-Kirchhoff stress tensor $\mathbf{S}_{UT}$, coming from the tensile test of the material, is $S_{UT11} = [\mathbf{S}_{UT}]_{11} = {}^S\sigma_y = \sqrt{r^2 - [a \cdot e^{pl} - \text{center}]^2}$. The equation defines an arc of a circle using three material parameters: the constant yield stress of the material σ_y, the maximum hardening stress Q by which the material can harden and the maximum accumulated plastic strain value $b = e^{pl}_{\max}$, at which the material loses its integrity, i.e. ${}^S\sigma_y = 0$. The relationship between the corresponding stress measures then can be written in tensor form as $\mathbf{P}_{UT} = \mathbf{F}_{UT} \cdot \mathbf{S}_{UT}$, in which the parameters σ_y, Q are second Piola-Kirchhoff stresses and $e^{pl} \subset \langle 0, b \rangle$. It should also be noted that the definitions of the accumulated plastic strain rate $\dot{e}^{pl}$ (an equivalent strain rate defined by Eq. $(67)_1$); the accumulated plastic strain e^{pl} (Eq. $(67)_2$), which controls the hardening and denotes the plastic damage; and the equivalent stress ${}^P\sigma_{eq}$ (Eq. $(63)_2$), respectively, which have the following physical meaning in the uniaxial tensile test of the material, $\dot{e}^{pl} = \partial^0 \dot{u}^{pl}_{UTX}/\partial X$, $e^{pl} = \partial^0 u^{pl}_{UTX}/\partial X$, ${}^P\sigma_{eq} = \mathrm{P}_{UT11} = F_{UT11} \cdot S_{UT11}$, have changed. The changes were required by consistency conditions, so that the model could work properly in either a one-dimensional (1D) stress state or a three-dimensional (3D) stress state, which may occur at a material particle during the analysis. In the above equations, the variables ${}^{spring}e^{pl}$ (see Eq. (65)) and ${}^{damper}e^{pl}$ (see Eq. (66)) denote the ductile and brittle damage defined in terms of the ratio of ductile and total damage increment xx [21].

2.5 Calculation of the plastic multiplier

A thermodynamically consistent formulation of the plastic flow allows for the calculation of the plastic multiplier in whatever (second Piola-Kirchhoff, first Piola-Kirchhoff, Kirchhoff and Cauchy) stress spaces using the corresponding definition of the yield surface. There are altogether two types of return mapping procedures for plastic multiplier calculation, which result in a thermodynamically consistent material model.

The first type of return mapping procedure, which is best suited for hyperelastic-plastic-based materials, is the ordinary return mapping procedure. Its thermodynamically consistent form in terms of the normality rules is defined by Eq. (38). Here the equation of the yield surface is solved directly for the plastic multiplier value, which in the case of our hyperelastic-plastic material takes the form (see Eq. $(63)_1$)

$$^{P}\sigma_{eq} - {^{P}\sigma_{y}} = 0. \tag{68}$$

In order to show that the return mapping is thermodynamically consistent, the equivalent stress in the material model is manipulated as follows:

$$^{P}\sigma_{eq} = \sqrt{\mathbf{P}:\mathbf{P}} = \frac{\mathbf{P}}{\sqrt{\mathbf{P}:\mathbf{P}}}:\mathbf{P} = \frac{\partial^{P}\Psi}{\partial \mathbf{P}}:\mathbf{P}. \tag{69}$$

Then after multiplying Eq. (68) by the plastic multiplier $\dot{\lambda}$ and the infinitesimal volume element dV_0 in the initial configuration of the body, the first term of Eq. (68) becomes the internal plastic power at a material constituent of the model $\dot{\lambda} \cdot {^{P}\sigma_{eq}} \cdot dV_0 = \dot{\lambda} \cdot \partial^{P}\Psi/\partial\mathbf{P} : \mathbf{P} \cdot dV_0 = \partial^0 \dot{\mathbf{u}}pl/\partial\mathbf{X} : \mathbf{P} \cdot dV_0 = d\dot{W}^{pl}$. Similarly, the second term of Eq. (68) becomes $\dot{\lambda} \cdot {^{P}\sigma_{y}} \cdot dV_0 = \dot{e}^{pl} \cdot {^{P}\sigma_{y}} \cdot S_0 \cdot dX = \partial^0 \dot{u}^{pl}_{UTX}/\partial X \cdot {^{P}\sigma_{y}} \cdot S_0 \cdot dX = d\dot{W}^{pl}$, which is just the internal plastic power at a material constituent of the specimen coming from the tensile test of the material, where $S_0 = S_0(X)$ is the cross-sectional area of the specimen.

The second type of return mapping procedure, which is best suited for hypoelastic-plastic-based materials, is the rate form of the return mapping procedure. Its thermodynamically consistent form in terms of the normality rules is defined by Eq. (39). Here the objective time derivative of the yield surface is used for the plastic multiplier calculation, which in the case of our hypoelastic-plastic material model takes the form

$$\frac{\partial^{P}\Psi}{\partial \mathbf{P}} : \mathfrak{L}_{\mathrm{P}}(\mathbf{P}) - [\mathfrak{L}_{\mathrm{P}}(\mathbf{P}_{UT})]_{11} = 0, \tag{70}$$

where $\mathfrak{L}_{\mathrm{P}}(\mathbf{P})$ is then replaced by the rate form of the constitutive equation of the material Eq. (43) and the second term on the LHS of Eq. (70) by the form $[\mathfrak{L}_{\mathrm{P}}(\mathbf{P}_{UT})]_{11} = F_{UT11} \cdot \left\{ \left[-a \cdot \left(a \cdot e^{pl} - \text{center} \right) \right] / \sqrt{r^2 - \left[a \cdot e^{pl} - \text{center} \right]^2} \right\} \cdot \dot{e}^{pl}$. It should be noted that the first term of Eq. (70) can be replaced by any other term of Eq. (39), because the formulation is thermodynamically consistent. Moreover, multiplying Eq. (70) by the plastic multiplier $\dot{\lambda}$ and the infinitesimal volume element dV_0 in the initial configuration of the body, it is easy to see that by solving Eq. (70), the following conservation equation is enforced:

$$\frac{\partial^0 \dot{\mathbf{u}}^{pl}}{\partial \mathbf{X}} : \mathfrak{L}_{\mathrm{P}}(\mathbf{P}) \cdot dV_0 - \frac{\partial^0 \dot{u}^{pl}_{UTX}}{\partial X} \cdot [\mathfrak{L}_{\mathrm{P}}(\mathbf{P}_{UT})]_{11} \cdot S_0 \cdot dX = 0. \tag{71}$$

Therefore, both return mapping algorithms in the above result in such a plastic multiplier calculation, during which the internal plastic power density of the models becomes the same as the internal plastic power density of the specimen, coming from the uniaxial tensile test of the material. It should also be noted that Eq. (70) is just the objective and thermodynamically consistent rate form of the Eq. (68) resulting from the invariance of the internal mechanical power and its first derivative Eq. (27), which now defines not only conjugate pairs of stress measures and strain or deformation rates but also conjugate differentiation operators and derivatives. Moreover, using Eqs. (20)–(23) and (40) and considering the fact that the physical meaning of Eqs. (68)–(71) is the conservation of plastic energy, the overall power at a material particle of the body dW can be decomposed into an elastic part dW^{el} and a plastic part dW^{pl} as follows:

$$dW = \dot{\mathbf{E}} : \mathbf{S} \cdot dV_0 = \frac{\partial^0 \dot{\mathbf{u}}}{\partial \mathbf{X}} : \mathbf{P} \cdot dV_0 = \mathbf{d} : \boldsymbol{\tau} \cdot dV_0 = \mathbf{d} : \boldsymbol{\sigma} \cdot dv = dW^{el} + dW^{pl}, \quad (72)$$

where

$$dW^{el} = \dot{\mathbf{E}}^{el} : \mathbf{S} \cdot dV_0 = \frac{\partial^0 \dot{\mathbf{u}}^{el}}{\partial \mathbf{X}} : \mathbf{P} \cdot dV_0 = \mathbf{d}^{el} : \boldsymbol{\tau} \cdot dV_0 = \mathbf{d}^{el} : \boldsymbol{\sigma} \cdot dv, \quad (73)$$

$$dW^{pl} = \dot{\lambda} \cdot \frac{\partial^S \Psi}{\partial \mathbf{S}} : \mathbf{S} \cdot dV_0 = \dot{\lambda} \cdot \frac{\partial^P \Psi}{\partial \mathbf{P}} : \mathbf{P} \cdot dV_0 = \dot{\lambda} \cdot \frac{\partial^\tau \Psi}{\partial \boldsymbol{\tau}} : \boldsymbol{\tau} \cdot dV_0 = \dot{\lambda} \cdot \frac{\partial^\sigma \Psi}{\partial \boldsymbol{\sigma}} : \boldsymbol{\sigma} \cdot dv, \quad (74)$$

which also proves that the formulation of the material model is thermodynamically consistent.

2.6 The ratio of ductile and total damage increment

The idea of the ratio of ductile and total damage increment xx was first introduced by Écsi and Élesztős in order to take into account the internal damping properly during plastic deformation of the hypoelastic-plastic material, where xx allowed for the redistribution of the plastic flow between the spring and the damper of a 1D frictional device representing the rheological model of the material [21]. The ratio is determined in an elastic corrector phase during return mapping, and its value is then kept constant. Since the return mapping procedure in our material model is carried out in the first Piola-Kirchhoff stress space, we had to modify the definition of the ratio as follows:

$$xx = \frac{\left\langle \mathbf{N} : \mathbf{F} \cdot \left({}^{mat}\mathbb{C}^{el} : \dot{\mathbf{E}} \right) \right\rangle}{\left\langle \mathbf{N} : \mathbf{F} \cdot \left({}^{mat}\mathbb{C}^{el} : \dot{\mathbf{E}} \right) \right\rangle + \left\langle \mathbf{N} : \mathbf{F} \cdot \left({}^{mat}\mathbb{C}^{vis} : \ddot{\mathbf{E}} \right) \right\rangle}, \quad (75)$$

where

$$\frac{\partial^P \Psi}{\partial \mathbf{P}} = \mathbf{N}, \qquad \mathbf{N} = \frac{\mathbf{P}}{\sqrt{\mathbf{P} : \mathbf{P}}} = \frac{\mathbf{P}}{\|\mathbf{P}\|}, \qquad \langle y \rangle = \frac{y + |y|}{2} \geq 0. \quad (76)$$

Eqs. (75) and (76) $\langle y \rangle$ denote the McCauly's brackets, which return zero if $y < 0$ and where we also used the transformation $\mathfrak{L}_P(\mathbf{P}) = \mathbf{F} \cdot \dot{\mathbf{S}}$. Please also note that all terms of Eq. (75) are objective stress rates, so that the value of xx is not affected by the change of the observer.

3. Numerical experiment

In our numerical experiment a cantilever, size 50 mm × 50 mm × 600 mm was studied applying dynamic pressure on 1/3 of its upper surface near the cantilever free end. The loading was defined as a product of a constant $p = 3$ MPa pressure and the Heaviside step function. The analysis was run as dynamic using $\Delta t = 5 \cdot 10^{-6}$ s time step size. **Table 1** lists the used material properties, and **Figure 2** depicts the spatially discretized model of the cantilever.

	Hypoelastic-plastic material with damping	Hyperelastic-plastic material with/ without damping
E [Pa]	7.31×10^{10}	7.31×10^{10}
E^{vis} [Pa · s]	7.31×10^{2}	$7.31 \times 10^{2}/0.0$
$\nu = \nu^{vis}$ [-]	0.33	0.33
σ_y [Pa]	220.0×10^{6}	220.0×10^{6}
Q [Pa]	110.0×10^{6}	110.0×10^{6}
b [-]	10^{64}	10^{64}
ρ_0 [kg · m^{-3}]	2770.0	2770.0

Table 1.
Material properties.

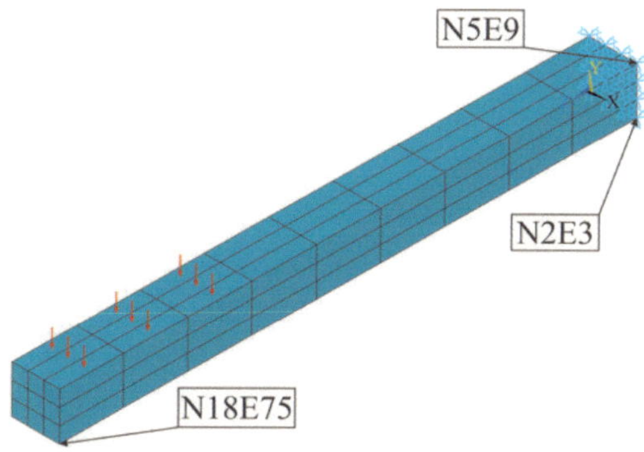

Figure 2.
The spatially discretized model.

In order to assess the value of the axial component of the deformation gradient coming from the tensile stress of the material F_{UT11}, we solved the one-dimensional (1D) rate form of the constitutive equation of the material (Eq. (42)) for the unknown component of the derivative of the axial elastic displacement field with respect to the axial material coordinate $\partial^0 u_x^{el}/\partial X$. The rate form of the constitutive equation of this specific 1D stress analysis, after neglecting the internal damping in the material, can be expressed in the following finite-strain form:

$$ {}^S\dot{\sigma}_y = \dot{S}_{11} = E \cdot \left[\frac{\partial^0 \dot{u}_x^{el}}{\partial X} \cdot \left(1 + \frac{\partial^0 u_x^{el}}{\partial X} + \frac{\partial^0 u_x^{pl}}{\partial X} \right) \right], \tag{77} $$

where ${}^S\dot{\sigma}_y = \dot{S}_{11}$ is the axial component of the second Piola-Kirchhoff stress rate tensor from the tensile test of the material and E is the Young's modulus. Furthermore considering that the accumulated plastic strain rate Eq. (67) $_1$ in this 1D stress state is $\dot{e}^{pl} = \partial^0 \dot{u}_x^{pl}/\partial X = \dot{\lambda}$, and that its integral is $e^{pl} = \partial^0 u_x^{pl}/\partial X$ (Eq. $(67)_2$), one can find F_{UT11} as a function of the accumulated plastic strain e^{pl} only in the following form:

$$ F_{UT11} = 1 + \frac{\partial^0 u_x^{el}}{\partial X} + \frac{\partial^0 u_x^{pl}}{\partial X} = 1 + \left[-\left(1 + e^{pl}\right) + \sqrt{\left(1 + e^{pl}\right)^2 + 2 \cdot \frac{{}^S\sigma y}{E}} \right] + e^{pl}, \tag{78} $$

where ${}^S\sigma y = {}^S\sigma y\left(e^{pl}\right)$ see also Eq. $(64)_1$.

4. Numerical results

Figure 3 depicts a few selected analysis results. These are the first principal Cauchy stress and the accumulated plastic strain distributions over the body, which is similar in the case of both materials, the vertical displacement time history curves and the accumulated plastic strain time history curves for the used material at selected nodes (see **Figure 2** for the location of the nodes).

It should be noted that in order to avoid problems with convergence with the used hypoelastic-plastic material, the value of the b parameter had to be set extraordinarily high. As a result, the isotropic hardening curve becomes flat in the range of the accumulated plastic strain value that occurred in the analysis, i.e. no isotropic hardening took place in the analysis. Moreover, for the same reasons, a very small viscous damping parameter has to be used with the model. The viscous damping, however, did not affect the analysis result at this kind of intensive loading, resulting in the plastic collapse of the beam, which can be seen in the time history curves. The analysis results otherwise seem to be reasonable.

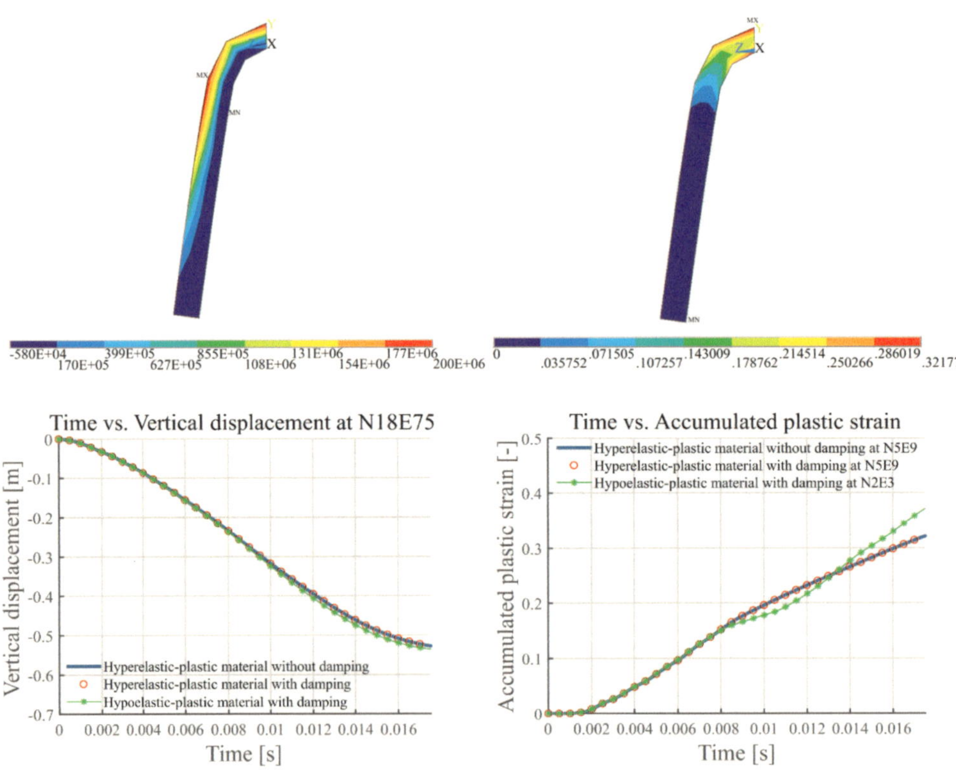

Figure 3.
Selected results from the analyses.

5. Conclusions

In this chapter an alternative framework for developing objective and thermodynamically consistent hypoelastic-plastic- and hyperelastic-plastic-based material models was presented using the first nonlinear continuum theory for finite deformations of elastoplastic media. The related material models were demonstrated in numerical experiments. The most important implication of the presented theory is that the analysis results of the related models are no longer affected by the description and the particularities of the mathematical formulation. The nonlinear continuum theory was also briefly presented, while the thermodynamic consistency of the formulation was in detail discussed. Another important implication of the theory is that the dissipated plastic power density of the model can directly be related to the dissipated plastic power density of the specimen coming from the uniaxial tensile stress of the modelled material. Moreover, contemporary tensile testing for material parameter determination will also have to be extended by determination of the deformation gradient of the specimen of the modelled material, as it is an important entry for the presented material models.

Acknowledgements

Funding from the VEGA 1/0740/16 grant, the KEGA 017STU-4/2018 grant and the APVV-15-0757 grant resources are greatly appreciated.

Author details

Ladislav Écsi*, Pavel Élesztős, Róbert Jerábek, Roland Jančo and Branislav Hučko
Department of Applied Mechanics and Mechatronics, Faculty of Mechanical Engineering, Slovak University of Technology in Bratislava, Bratislava, Slovakia

*Address all correspondence to: ladislav.ecsi@stuba.sk

References

[1] Simo JC, Hughes TJR. Computational Inelasticity. New York: Springer; 2000. 392 p. ISBN-10: 0387975209

[2] Nemat-Nasser S. Plasticity: A Treatise on Finite Deformation of Heterogeneous Inelastic Materials. Cambridge: Cambridge University Press; 2004. 760 p. ISBN-10: 0521108063

[3] Belytschko T, Liu WK, Moran B. Nonlinear Finite Elements for Continua and Structures. 2nd ed. Chichester: Wiley; 2014. 830 p. ISBN-10: 1118632702

[4] De Souza Neto EA, Perić D, Owen DRJ. Computational Methods for Plasticity: Theory and Applications. 1st ed. Singapore: Wiley; 2008. 814 p. ISBN-10: 0470694521

[5] Asaro RJ. Micromechanics of crystals and polycrystals. In: Hutchinson JW, Wu TY, editors. Advances in Applied Mechanics. Vol. 23. New York: Academic Press; 1983. pp. 1-115. ISBN-10: 0120020238

[6] Peirce D, Asaro RJ, Needlemann A. An analysis of nonuniform and localized deformation in ductile single crystals. Acta Metallurgica. 1982;**30**:1087-1119. DOI: 10.1016/0001-6160(82)90005-0

[7] Peirce D. Shear band bifurcation in ductile single crystals. Journal of the Mechanics and Physics of Solids. 1983; **31**:133-153. DOI: 10.1016/0022-5096(83)90047-9

[8] Kojić M, Bathe KJ. Studies of finite element procedures–Stress solution of a closed elastic strain paths with stretching and shearing using the updated Lagrangian Jaumann formulation. Computers and Structures. 1987;**28**(1/2):175-179. DOI: 10.1016/0045-7949(87)90247–1

[9] Meyers A, Xiao H, Bruhns OT. Choice of objective rate in single parameter hypoelastic deformation cycles. Computers and Structures. 2006; **84**:1134-1140. DOI: 10.1016/j.compstruc.2006.01.012

[10] Lehmann T. Anisotrope plastishe Formänderungen. Romanian Journal of Technical Sciences. Series: Applied Mechanics. 1972;**17**:1077-1086. DOI: 10.1007/BF02096162

[11] Diens JK. On the analysis of rotation and stress rate in deforming bodies. Acta Mechanica. 1979;**32**:217-232. DOI: 10.1007/BF01379008

[12] Simo JC, Pistner KS. Remark on the rate of constitutive equations for finite deformation problems: Computational implications. Computer Methods in Applied Mechanics. 1984;**46**:201-215

[13] Perić D. On constitutive stress rates in solid mechanics: Computational implications. The International Journal for Numerical Methods in Engineering. 1992;**33**:799-817. DOI: 10.1002/nme.1620330409

[14] Écsi L, Élesztős P. An alternative material model using a generalized J2 finite-strain flow plasticity theory with isotropic hardening. International Journal of Applied Mechanics and Engineering. 2018;**2**(23):351-365. DOI: 10.2478/ijame-2018-0019

[15] Gurtin ME. An Introduction to Continuum Mechanics. Orlando: Academic Press; 1981. p. 265. ISBN-10: 9780123097507

[16] Holzapfel GA. Nonlinear Solid Mechanics. A Continuum Approach for Engineering. Chichester: Wiley; 2001. p. 455. ISBN-10: 0471823198

[17] Bonet J, Wood RD. Nonlinear Continuum Mechanics for Finite

Element Analysis. 2nd ed. Cambridge: Cambridge University Press; 2008. p. 340. ISBN-10: 0521838703

[18] Spencer AJM. Continuum Mechanics. 1st ed. New York: Longman; 2012. p. 208. ISBN-10: 0486435946

[19] Borisenko AI, Tarapov IE. Vector and Tensor Analysis with Applications. New York: Dover; 2012. p. 288. ISBN-10: 0486638332

[20] Maugin GA. The saga of internal variables of state in continuum thermo-mechanics. Research Communications. 2015;**69**:79-86. DOI: 10.1016/j.mechrescom.2015.06.009

[21] Écsi L, Élesztős P. Moving toward a more realistic material model of a ductile material with failure mode transition. Materialwissenschaft und Werkstofftechnik. 2012;**5**(43):379-387. DOI: 10.1002/mawe.201200969

Section 4

Applications for Nuclear Fuel Elements

Chapter 5

CERMETS for Use in Nuclear Thermal Propulsion

Dennis S. Tucker

Abstract

NASA is currently investigating nuclear thermal propulsion as an alternative to chemical propulsion for manned missions to the outer planets. There are a number of materials being considered for use as fuel elements. These materials include tricarbides and CERMETS such as W/UO_2, Mo/UO_2, W/UN and Mo/UN. All of these materials require high temperature processing to achieve the required densities. It has been found that Spark Plasma Sintering is a good choice for sintering these materials to the required densities while maintaining a uniform grain size. In this chapter a brief history of NASA's research into nuclear thermal propulsion will be given, followed by specific research by this author and others to produce CERMET fuels.

Keywords: nuclear thermal propulsion, fuel elements, spark plasma sintering, density, uranium distribution

1. Introduction

Long duration spaceflight can expose astronauts to two major problems. These are extended periods of weightlessness and radiation exposure. Thus, it is necessary to develop alternate means of propulsion to that of chemical propulsion. A reasonable option being studied by NASA is nuclear thermal propulsion (NTP), whereby nuclear fuel elements made from metal/UO_2, metal/UN or tricarbides are being considered. These fuel elements must be capable of operating in excess of 2700 K while being compatible with the propellant, typically hydrogen [1]. From previous studies it was shown that W/UO_2 as a fuel element can be used both in power and propulsion at temperatures as high as 3000 K [2–4]. Two factors need to be considered when producing W/UO_2 fuel elements. The first is the importance of a uniform distribution of UO_2 particles in the tungsten matrix. If one has segregation of UO_2 particles it can lead to hot spots and ultimately failure of the fuel element. The second factor which needs to be considered is the stoichiometry of the UO_2 particles. Maintaining stoichiometry is vital to ensure stability and proper operation of the fuel element. This chapter will detail a brief history of NTP by NASA including fuel element work, followed by more recent research on various fuel systems under consideration.

IntechOpen

2. NTP history

As far back as the 1940s, it was recognized that energy from nuclear fission could be used to power spacecraft by heating a working fluid such as hydrogen and provide thrust to the rocket via expansion of the propellant through a rocket nozzle. A simple drawing of an NTP rocket is shown in **Figure 1** below.

Due to the high specific impulse, NTP is considered to be the preferred propulsion method for future manned flights to Mars. Specific impulse (I_{sp}) is a method to measure and compare the efficiency of different propulsion systems. It is determined by the ratio of thrust to the propellant mass flow rate through the engine. The typical NTP engine would have and I_{sp} = 900 s, which is twice that of chemical propulsion systems. During a Mars manned mission the engine would be run for a total of 4 hours. One hour to accelerate from Earth to Mars, followed by a 1 hour deceleration burn. The same burn cycles would occur on a return trip. Thus it is quite critical for the fuel elements to retain their stability during these burns.

The United States was involved in the production and testing of NTP engines during the period of 1955–1972. This was the Rover/NERVA program which tested 20 prototype reactors during this period. These prototypes included fuel test reactors, a safety reactor and prototype engines. **Figure 2** below shows a test of one of the NERVA engines. This engine reached an Isp of ~850 s during a 2 hour burn. Twenty prototype reactors were ground tested. Fuel forms evolved over the duration of the program [2–9].

The fuel elements used during testing were of varying compositions. These were coated graphite-matrix elements followed by advanced fuel elements consisting of UC-ZrC-C and all carbide elements ((U, Zr)C) [5–8]. Most of the testing was performed using the coated graphite fuel elements. These elements were full length (52″) with a hexagonal cross section (0.75″ flat-to-flat) and 19 axial holes for propellant flow. These elements were arranged to create a cylindrical reactor core. The NERVA/Rover program proved NTP to be a viable technology [9]. Several prototype reactors were produced which survived multiple restarts and power levels over 4000 MW, thrust levels of 250 klbf, maximum propellant outlet temperatures of 2550 K, a maximum net specific impulse of 850 s and over an hour of

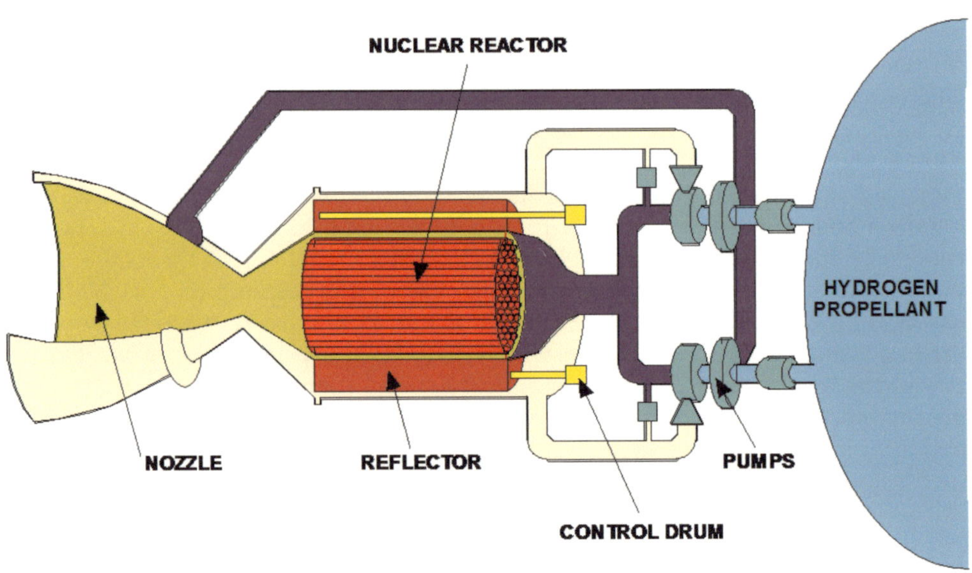

Figure 1.
Schematic of NTP rocket.

Figure 2.
Test of NERVA engine.

continuous operation [9]. The fuel elements had to perform various tasks. The elements contained fissile material (UO_2 or UC_2) and graphite as a moderator. The fuel elements also functioned as structural components.

3. Present-day NTP fuels research

Presently there are a number of fuels under consideration for NTP. These include graphite composites, tricarbides (U-Zr-Nb)C and CERMETS (MUO_2 and MUN). At Marshall Space Flight Center, we are concentrating on CERMETS (W/UO_2, Mo/UO_2, W/UN, Mo/UN) and tricarbides. Consolidation of these

CERMETS and tricarbides has been performed using RF induction furnaces, hot pressing and spark plasma sintering (SPS).

With respect to CERMET processing a number of approaches have been utilized in order to obtain a uniform distribution of the fissile material within the metal matrix. These include traditional powder processing techniques and coating the fissile material with metal using chemical vapor deposition (CVD) [10]. Recently, a fluidized bed reactor to coat UO_2 with tungsten was tested [11]. CVD shows great promise, however, there are technical issues due to the complexity of the experimental CVD apparatus and the CVD process. These issues lead to expense and long reaction times to apply the appropriate thickness coatings. Thus, a technique for obtaining a uniformly coated spherical UO_2 particles was developed [12]. In this technique a small amount of high density polyethylene binder (0.25 w/o) is added to a mixture of tungsten and UO_2 particles. The powders are then mixed thoroughly using a turbula, then heated and mixed on a magnetic stir plate.

Traditional sintering methods can be used to densify W/UO_2. Both hot pressing and hot isostatic pressing have been used. There are drawbacks to these two methods including incomplete sintering and dissociation of UO_2 at high temperatures, pressures and long sintering times. Another issue is a problem of exaggerated grain growth which can occur under these processing conditions.

A sintering method which has been shown to be a reasonable alternative to these traditional method is Spark Plasma Sintering (SPS) [13–15]. SPS leads to higher densities at lower temperatures and processing times while minimizing grain growth. Grain growth is detrimental to densification during the sintering process. In one study, UO_2 was produced by hot pressing, however it was found that a large number of pores were present on the grain faces which led to a smaller grain boundary contact area [16]. In this same study it was observed that grains which had not undergone exaggerated grain growth had pores at the grain corners. It has been observed that pores located on grain faces have greater mobility than those at grain corners and ultimately end up within the grains [17, 18]. Joule heating is utilized in SPS which results in passing a current through the powder during sintering [19]. A pulsed current is utilized in SPS which leads to two different operating temperatures: the average temperature and the maximum temperature. The average temperature is lower than the melting point of the materials. During current discharge, material is transported by a plasma across pores of the matrix. While the pulse is off, the matrix cools rapidly, and this lead to condensation of the material vapor within regions where there is mechanical contact between grains. This mechanism leads to necking between grains. There have been a number of studies have using SPS to consolidate tungsten and a surrogate or tungsten and UO_2. In one study W/CeO_2 was sintered using SPS [19]. In a second study W/UO_2 was densified using SPS [20]. A shortcoming in both of these studies was the segregation of the tungsten and the oxides. In the W/UO_2 study, the materials were mixed in a turbula for 1 hour then hot isostatic pressed [21]. This result was a segregated CERMET due to the differences in powder sizes (W-15 μm, UO_2-200 μm) and density differences where size differences made the largest difference. In **Figure 3** on can plainly see segregation in the sintered CERMET.

Studies [22, 23] were undertaken to eliminate this segregation using an inexpensive, simple technique. Depleted UO_2 particles were obtained from Oak Ridge National Laboratory. These particles had an average size of 200 μm. Tungsten powder with a particle size of 5/15 μm was purchased and used as the matrix material. In order to coat the UO_2 particles with tungsten, a powder processing technique was developed. In this technique, high molecular weight polyethylene powder was milled to approximately 1 μm in diameter. Next a mixture of 50 g of 60 vol% UO_2, 40 vol% W and 0.25 wt% polyethylene powder were thoroughly

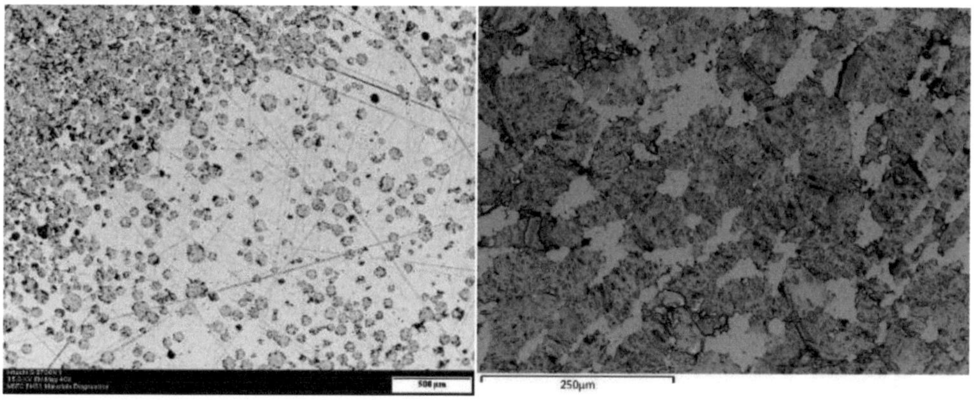

Figure 3.
Scanning electron micrographs of hot isostatic pressed W/UO_2. Dark area on left is UO_2 while on right shows higher magnification image of UO_2.

mixed for 45 minutes in a turbula. This powder was placed in a 400 ml Pyrex beaker and then stirred on a hot plate for 10 minutes above the drop point of the polyethylene (101°C). This process was repeated until 500 g was produced. The powder was shipped to the Center for Space Nuclear Research in Idaho Falls, Idaho for sintering in the SPS. Thirty one grams was placed in a graphene die for sintering. Samples were densified at 1600, 1700, 1750, 1800 and 1850°C. Samples were heated at a rate of 100°C/minute to the sintering temperature. The pressure was increased by 10 Mpa/minute to 50 Mpa. After soaking at the maximum temperature for 20 minutes, the pressure was decreased by 10 Mpa/minute to 5 Mpa and the temperature was decreased by 20°C/minute to room temperature.

Density was obtained using the Archimedes method. Carbon content was analyzed using instrumental gas analysis (EAG, NY). Scanning electron microscopy with energy dispersive x-ray analysis was performed on all samples. Microstructural and chemical analyses were carried out by using transmission electron microscopy (TEM) and atom probe tomography (APT) techniques. TEM and APT specimens were prepared at phase boundaries using lift-out methods with a focused ion beam (FIB). The size of each TEM lamella was 10 × 10 μm. TEM characterization was carried out using a FEI Tecnai G2 F30 STEM FEG equipped with energy dispersive x-ray spectrometry (EDS). The EDS analyses were done in Scanning TEM (STEM) mode with a beam size of 1 nm. APT was carried out using a CAMECA LEAP 4000X HR. APT data reconstruction was done using the CAMECA IVAS software.

Figure 4 shows a scanning electron micrograph of UO_2 particles coated with tungsten powder.

In **Figure 4** one can note that the UO_2 particles are almost completely covered with the tungsten powder. The polyethylene binder is viscous above its drop point (101°C) and coats both the UO_2 particle and tungsten particles which when subsequently mixed together results in the image in **Figure 4**. The mixing temperature was 140°C which led to a binder viscosity of 140 cP. As the mixture was stirred, the nearly spherical UO_2 particles rolled around the bottom of the beaker and were coated with the tungsten particles.

As can be seen in **Table 1** below, the density is relatively high at a sintering temperature of 1600°C and gradually increases up to 1800°C and jumps to 99.46% of theoretical density at 1850°C.

The lower sintering temperature densities align well with what was previously reported for which the density was reported as 97.9% of theoretical for W-Re/UO_2 at 1500°C and 40 Mpa applied pressure using SPS [20]. The higher sintering

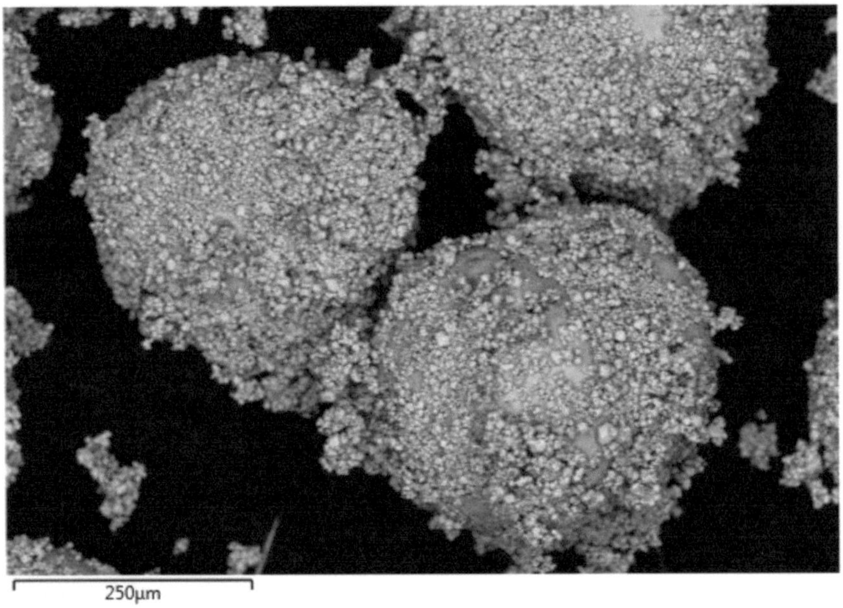

Figure 4.
Scanning electron micrograph showing UO_2 particles coated with tungsten powder.

Sintering Temperature °C	Density, % of Theoretical
1600	97.18
1700	98.19
1750	98.70
1800	98.57
1850	99.46

Table 1.
Sintering temperature versus % theoretical density for SPS W/UO_2.

temperatures and increased applied pressure used in this study can account for an increase in density seen in **Table 1**. It is also thought that decreasing the cooling rate to 20°C/minute could also have contributed to further densification. At lower cooling rated the insulated die will retain heat which will allow more densification.

Figure 5 below is a low magnification SEM micrograph of the sample sintered at 1850°C.

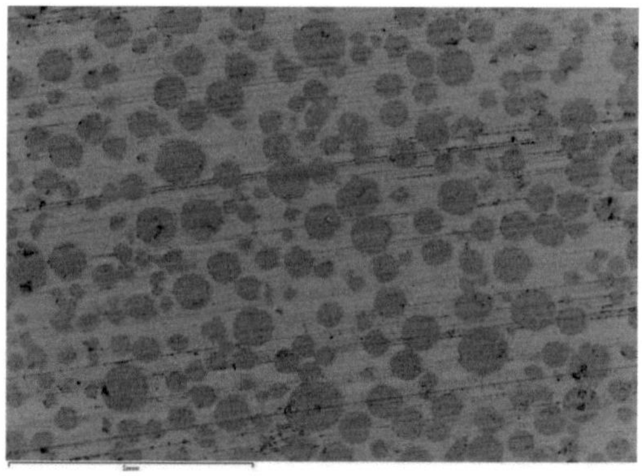

Figure 5.
SEM of SPS sintered W/UO_2 at 1850°C.

This figure shows the distribution of UO_2 particles within the tungsten matrix. The UO_2 particles are the darker near spherical particles in the lighter gray tungsten matrix. As can be seen, the distribution of UO_2 particles is nearly uniform with the tungsten matrix. Obtaining a uniform mixture of disparate particle sizes is extremely difficult using traditional powder mixing techniques without the aid of a binder. There four properties which give rise to segregation in powder mixes are: particle size difference, variations in particle density, along with shape and resilience [24]. Particle size difference has been shown to be the most important factor [21]. Segregation of powders causes fluctuations in the size distributions of particles and this leads to variations in bulk density which can affect the desired properties. There are three mechanisms of segregation which can occur during mixing and vibration. Vibration is often used to increase packing density in the powder which in leads to higher sintered densities. Segregation can occur during mixing when fine particles travel further than coarse particles during the mixing operation. If a mass of particles is disturbed so that individual particles move, a rearrangement can take place. This is termed percolation. Over time, gaps between particles occur, which allows particles from above to move down, while a particles from some other location replaces them. When the powder mass contains different size particles, small particles will fall through the large particle interstices leading to segregation. Percolation occurs when the mass of particles is disturbed due to a shear stress within the particle mass. This phenomenon is explained in which a large particle causes an increase in pressure in the region below it which compacts the material and stops the particle from moving downward. Upward movement allows fines to run in under the coarse particle and these in turn lock in position. If the vibration intensity is large enough the larger particles will migrate to the powder surface. The powder process described above which uses polyethylene binder, overcomes this difficulty using a minimum amount of binder which burns out during the SPS process and is drawn away from the sample by the vacuum system. In all the sintering temperatures listed in **Table 1** the vacuum was $\sim 2 \times 10^{-3}$ Torr. The binder acts as an adhesive for the W/UO_2 powders eliminating the disparate particle size effect. It was found that the carbon content for the mixed powders was 0.025 wt%, while the sintered samples were below the detectable limit which is in parts per million.

Figure 6 below shows EDS maps for the sample sintered at 1850°C.

In this figure one can see the SEM in the figure on the left and the x-ray maps on the center and right figure. The UO_2 particles are blue and the tungsten matrix is orange. One can also note some fracturing of the UO_2 particles. This is most likely due to the pressure during sintering but could also be caused during the grinding and polishing operation. This should be avoided to lessen the probability that uranium can escape and diffuse to the tungsten matrix and to the fuel surface. One can also note some UO_2 particle-particle contact which can lead to hot spots in the fuel during operation. This in turn could lead to disruption of the fuel element.

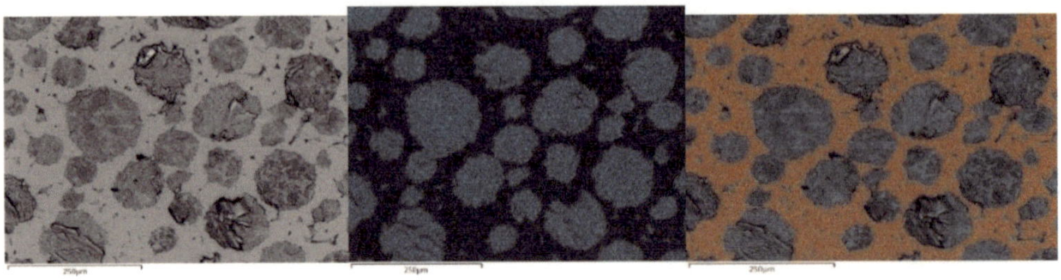

Figure 6.
SEM (left figure), EDS maps (UO_2—blue and W—orange) for sample sintered at 1850°C.

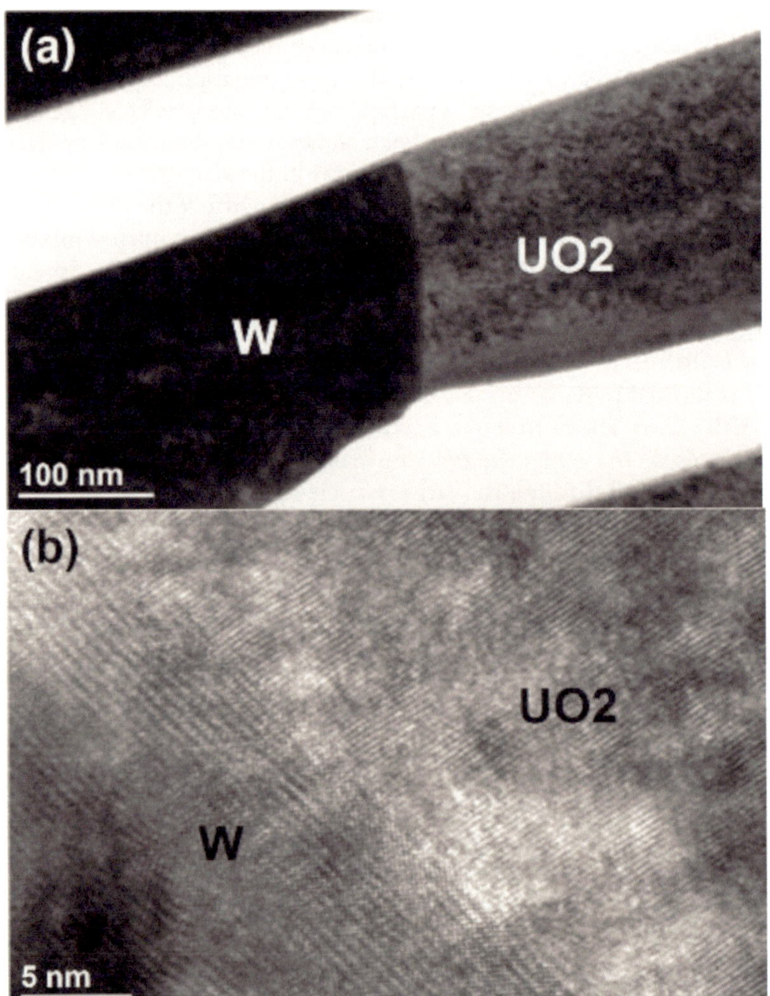

Figure 7.
HRTEM image of boundary of W/UO_2 for sample sintered at 1850°C: (a) is the low resolution boundary and (b) is the high resolution image of this boundary.

Using high resolution transmission electron microscopy (HRTEM) it is possible to image the boundaries between UO_2 particles and the tungsten matrix. **Figure** 7 below shows a typical region in the sample sintered at 1850°C.

This type of boundary is typical for all samples except for the one sintered at 1600°C which also showed an anomalous third phase. This is shown in **Figure 8** below.

There was an anomalous phase which was identified as $U_{0.1}WO_3$, space group Pm-3m (221) which is a cubic structure. This was based on its electron diffraction pattern and by considering the atomic ratio of U:W = 1:10 which is consistent with EDS results. Tungsten trioxide, WO_3, has a monoclinic structure with a space group mP32. It is possible that this anomalous phase is WO_3 with uranium contamination, since $U_{0.1}WO_3$ phase has not been previously identified in the literature. Since the space group and crystal structure are different for these two phases, this could be a new phase. High resolution, high intensity x-ray diffraction could be performed on all sintered samples to make a definite determination. It could be that the $U_{0.1}WO_3$ phase forms due to the availability of oxygen vacancies from the UO_2 reduction due to sintering in vacuum. The EDS line scan across the W/UO_2 boundary for the sample sintered at 1850°C is shown in **Figure 9a**.

Figure 9a shows the length of the line scan across the boundary. In **Figure 9b** it can be seen that the uranium has diffused approximately 15 nm into the tungsten matrix. The green line is the uranium curve and one measures where it crosses over the blue curve (tungsten). For all other sintered samples, it was seen that the uranium diffused approximately 10 nm into the tungsten matrix. The atom imaging probe analysis for the sample sintered at 1850°C is shown below in **Figure 10**.

It can be seen from **Figure 10** that the uranium and oxygen were present in the form of UO, UO_2 and UO_3. The nitrogen present is most likely from the nitrogen gas backfill used during SPS at room temperature after cooling. Silicon was observed for all samples except the one processed at 1750°C. Its origin is unknown but most likely is an impurity picked up during grinding and polishing. The carbon present in the sample which is from the polyethylene binder used during powder processing. The above data led to a formula given as $UO_{1.95}$ which is slightly sub-stoichiometric. The samples sintered at 1600, 1650 and 1700°C were also calculated to have this same formula. The only difference was for the sample processed at 1750°C which had the formula UO_2.

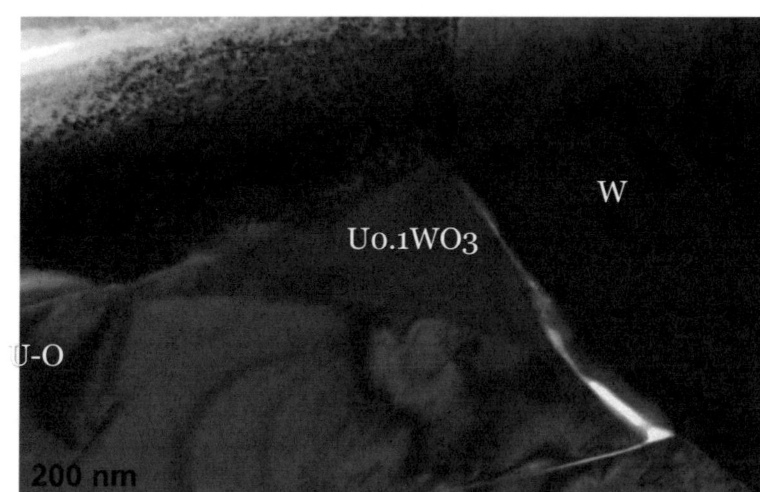

Figure 8.
TEM of sample sintered at 1600°C showing phase $U_{0.1}WO_3$.

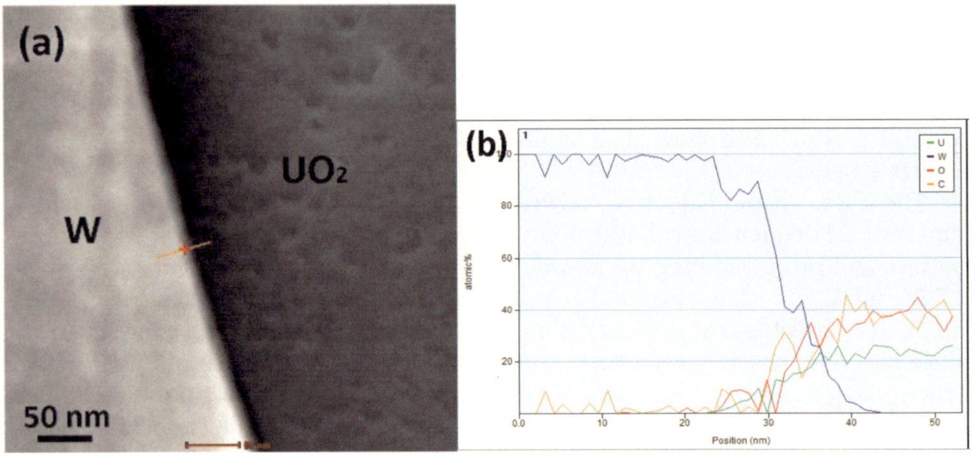

Figure 9.
EDS line scan across the W/UO_2 boundary for sample sintered at 1850°C in (a) and the results are in (b).

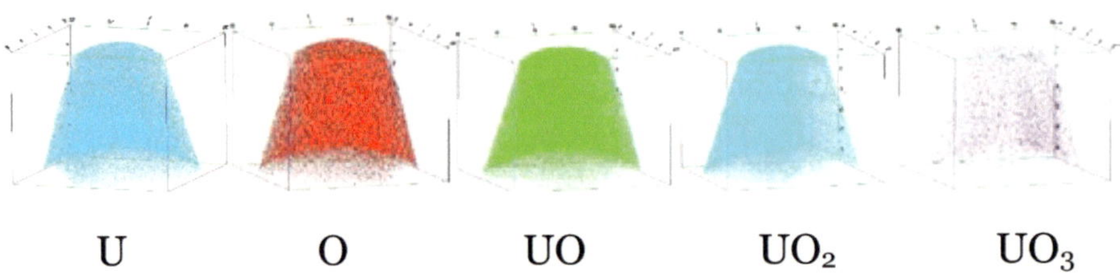

Ion	Atomic %	Atomic Error %
C	4.001599093	1.36E-03
N	0.01180554	1.27E-03
O	62.9657518	1.02E-02
U	32.15545232	1.17E-03
Si	0.40945344	4.80E-03

Figure 10.
3-D element maps from UO_2 particles and atomic percent from sample sintered at 1850°C.

The loss of uranium from the UO_2 particles and uranium migration into the tungsten matrix can be understood in terms of the generation of oxygen vacancies during sintering in a vacuum environment. An reaction for UO_2 if oxygen vacancies are abundant is given by Eq. (1).

$$UO_2 = UO_{2-x} + x/2O_2 \tag{1}$$

With the loss of oxygen there are two possible defect reactions that can occur. The first reaction is electronic compensation leading to the creation of oxygen vacancies and electrons. This is shown in Eq. (2).

$$O_o = V\ddot{o} + ½O_2 + 2e \tag{2}$$

Ionic substitution can lead to the formation of oxygen vacancies and reduction of the metal oxide on their sites as shown below in Eq. (3).

$$O_o = V\ddot{o} + 1/2O_2 + 2\,M'_M \quad (3)$$

The result of either of these reactions will be a sub-stoichiometric uranium oxide and free uranium as shown in Eq. (4).

$$2UO_{2-x} \rightarrow (2-x)UO_2 + xU \quad (4)$$

The free uranium from this reaction is then available to diffuse into the tungsten matrix. This mechanism occurs due to Fick's law of diffusion. The importance of the presence of free uranium in sintered W/UO_2 samples cannot be overstated. These materials will be exposed to hydrogen gas in a thermal cycling environment during engine operation. When thermal cycling takes place in a hydrogen environment, hydrogen will penetrate into the tungsten matrix by both grain boundary and bulk diffusion. The hydrogen can then combine with the free uranium leading to uranium hydride. Uranium hydride can also be formed by reaction with the UO_2 particle. This is shown in Eqs. (5) and (6) below.

$$UO_2 + H_2 \rightarrow UO_{2-x} + xH_2O \quad (5)$$

$$2UO_{2-x} \rightarrow (2-x)UO_2 + xU \quad (6)$$

The free uranium has a melting point of ~1130°C and will rapidly diffuse along the tungsten grain boundaries and form UH_3 at ~225°C. The formation of UH_3 leads to large increases in volume and which can result in tungsten grain separation. This grain separation creates avenues for migration of UO_2 to the CERMET surface. This results in the loss of UO_2 and can lead to mechanical failure. The free uranium not only forms UH_3, but can also reoxidize to form UO_2. Both mechanisms result in a large volume expansion and loss of mechanical integrity. There is also a difference between isothermal and cyclic heating. It has been shown that cycling heating results in more fuel loss than isothermal heating in hydrogen [25].

It has been found that oxides such as ThO_2, Ce_2O_3 and Y_2O_3 reduce fuel loss when added to the CERMET powder [25]. The observation was that the oxide additives did not increase the solubility of uranium in UO_2, but stabilized UO_2 against oxygen loss. Two mechanisms were proposed to explain the stabilization against oxygen loss: (1) oxide additives lower the partial molar free energy of oxygen in the UO_2. This precludes the possibility of forming free uranium upon cooling and (2) when metal oxide is added to the CERMET powder, uranium is transformed to a hexavalent state. This hexavalent state precludes the formation of uranium metal. The UO_2 maintains an oxygen-to-metal ration of 2.0–2.1 by forming a defect lattice structure. To maintain electrical neutrality, the uranium ions will be in the hexavalent state. U^{4+} cannot be reduced to the metal in the presence of U^{6+}. Thus, the initial loss of oxygen from the CERMET will be accompanied by oxygen vacancies rather than the formation of free uranium. The use of hyperstoichiometric uranium oxide (UO_{2+x}). UO_2 CERMETS in which the O/U ratio of the starting composition was varied between 1.93 and 2.05 was studied. It was shown in this study that there was minimal effect of varying this ratio. Thus the most likely candidate to stabilize UO_2 during sintering and thermal cycling in hydrogen will be a rare earth oxide addition.

4. Conclusions

In this chapter a brief history of the nuclear thermal propulsion program was given. A present-day research into processing and properties of nuclear fuel elements was discussed. In particular W/UO_2 which was spark plasma sintered was discussed. Uranium migration into the tungsten matrix was observed for all samples. The presence of uranium was explained in terms of oxygen vacancy generation due to processing in vacuum and the migration of the uranium by Fick's law of diffusion. Possible solutions to this problem were also discussed.

Acknowledgements

The author would like to thank the Nuclear Thermal Propulsion office at Marshall Space Flight Center for funding this work. The author would also like to thank the Center for Space Nuclear Research for performing the spark plasma experiments and the Center for Advanced Energy Studies for performing TEM and atom probe measurements. Both institutes are located in Idaho Falls, Idaho, USA.

Conflict of interest

There is no conflict of interest represented by this work.

Author details

Dennis S. Tucker
NASA Marshall Space Flight Center, Alabama, United States

*Address all correspondence to: dr.dennis.tucker@nasa.gov

References

[1] Burkes DE, Wachs DM, Werner JE, Howe SD. An Overview of Current and Past W-UO_2 CERMET Fuel Fabrication Technology. Space Nuclear Conference; 2007

[2] Bhattacharyya SK. An Assessment of Fuels for Nuclear Thermal Propulsion, ANL/TD/TM01-22; 2001

[3] Marlowe MO, Kaznoff AI. Development of Low Thermal Expansion W/UO_2 Nuclear Fuel, NASA-CR-72711; 1970

[4] Baker RJ. Basic Behavior and Properties of W/UO_2 CERMETS, NASA-CR-54840; 1965

[5] Koenig DR. Experience Gained from the Space Nuclear Rocket Program (ROVER), Rep. No. LA-10062-H; Los Alamos National Laboratory; 1986

[6] Lyon LL. Performance of (U,Zr)C-Graphite (Composite) and (U,Zr)C (Carbide) Fuel Elements in the Nuclear Furnace 1 Test Reactor, Rep. No. LA-5398-MS; Los Alamos National Laboratory; 1973

[7] Wallace TC. Review of Rover fuel element protective coating development at Los Alamos. In: Proceedings of the Eight Symposium on Space Nuclear Power Systems. 1991. pp. 1024-1036

[8] Taub JM. A Review of Fuel Element Development for Nuclear Rocket Engines, Rep. No. LA-5931; Los Alamos National Laboratory; 1975

[9] Qualls L. Graphite Fuel Development for Nuclear Thermal Propulsion; Oak Ridge National Laboratory; 2013

[10] Benensky K. Summary of historical solid Core nuclear thermal propulsion fuels [thesis]. State College Pennsylvania, Penn State University; 2013

[11] Hickman RR, Broadway JW, Mirales OR. Fabrication and testing of CERMET fuel materials for nuclear thermal propulsion. In: AIAA Joint Propulsion Conference; 13 July-1 August 2012; Atlanta, Georgia. 2012

[12] Tucker DS, O'Connor A, Hickman R. A methodology for producing a uniform mixture of UO_2 in a tungsten matrix. Journal of Physical Science and Applications. 2015;**5**:255-262. DOI: 10.17265/2159-5348/2015.04.002

[13] O'Brien RC, Ambrosi RM, Bannister NP, Howe SD, Atkinson HV. Spark plasma sintering of simulated radioisotope materials within tungsten CERMETS. Journal of Nuclear Materials. 2009;**393**:108-113. DOI: 10.1016/j.jnucmat.2009.05.012

[14] Wang X, Xie Y, Fuo H, Van der Blest O, Vieugels J. Sintering of WC-Co powder with nanocrystalline WC by spark plasma sintering. Rare Metals. 2006;**25**:246

[15] Atkinson HV, Davies S. Fundamental of hot isostatic pressing: An overview. Metallurgical and Materials Transactions. 2000;**A31**:2981

[16] Xu A, Soloman AA. The effects of grain growth on the intergranular porosity distribution in hot pressed and swelled UO_2. In: Proceedings of Ceramic Microstructures '86; July 1986; Plenum, Berkley, CA: University of California. 1986. pp. 28-31. ISBN: 0306426811

[17] Carpay FMA. The effect of pore drag on ceramic microstructures. In: Proceedings of Ceramic Microstructures '76; Westview, Boulder, CO. 1977. p. 171. ISBN: 0891583076

[18] German RM. Sintering Theory and Practice. New York: John Wiley; 1996. ISBN: 0-471-05786-X

[19] O'Brien RC, Ambrosi RM, Bannister NP, Howe SD, Atkinson HV. Safe thermoelectric generators and heat sources for space applications. Journal of Nuclear Materials. 2008;**377**:506

[20] O'Brien RC, Jerred ND. Spark Plasma Sintering of W-UO_2 CERMETS; 2013. pp. 433-450

[21] Williams JC. The segregation of particulate materials: A review. Powder Technology. 1976;**15**:245

[22] Tucker DS, Barnes MW, Hone L, Cook S. High density, uniformly distributed W/UO_2 for use in nuclear thermal propulsion. Journal of Nuclear Materials. 2017;**486**:246-249. DOI: 10.1016/jnucmat.2017.01.033

[23] Tucker DS, Wu Y, Burns J. Uranium migration in spark plasma sintered W/ UO_2 CERMETS. Journal of Nuclear Materials. 2018;**500**:141-144. DOI: 10.1016/jnucmat.2017.12.029

[24] Haertling C, Hanrahan RJ. Literature review of thermal and radiation performance parameters for high-temperature uranium dioxide fueled CERMET materials. Journal of Nuclear Materials. 2007;**366**:317-335. DOI: 10.1016/jnucmat.2007.03.024

[25] Saunders NT, Gluyas RE, Watson GK. Feasibility of a Tungsten-Water Mediated Rocket II. NASA Lewis Research Center Report, NASA-TM-X-1421; 1968